LONGING FOR THE HARMONIES

LONGING FOR THE HARMONIES

Themes and Variations from Modern Physics

Frank Wilczek and Betsy Devine

W·W·Norton & Company· New York· London

Copyright © 1988 by Frank Wilczek and Betsy Devine
All rights reserved.
Published simultaneously in Canada by Penguin Books Canada Ltd., 2801 John
Street, Markham, Ontario L3R 1B4.
Printed in the United States of America.
First Edition

The text of this book is composed in Century Schoolbook, with display type set in
Century Old Style. Composition and manufacturing by The Haddon Craftsmen, Inc.
Book design by Bernard Klein.

The lines from T. S. Eliot's poem "Little Gidding" are from *Four Quartets*, copyright
1936 by Harcourt Brace Jovanovich, Inc.; copyright © 1963, 1964 by T. S. Eliot.
Reprinted by permission of the publishers.

Library of Congress Cataloging-in-Publication Data

Wilczek, Frank.
Longing for the harmonies.

Includes index.
1. Physics. 2. Physics—Philosophy. 3. Harmony
(Philosophy) I. Devine, Betsy. II. Title.
QC21.2.W52 1987   530   87-7653

ISBN 0-393-02482-2

W. W. Norton & Company, Inc., 500 Fifth Avenue, New York, N. Y. 10110
W. W. Norton & Company Ltd., 37 Great Russell Street, London WC1B 3NU

1 2 3 4 5 6 7 8 9 0

To Amity and Mira:

*In vain does the God of War growl, snarl, roar, and try to interrupt
with bombards, trumpets, and his whole tarantaran. . . . Let us despise
the barbaric neighings which echo through these noble lands and
awaken our understanding and longing for the harmonies.*

—Kepler

Contents

iii. *Transformations*

iv. *Inevitability*

v. *Quantal Reality*

vi. *Radical Uniformity in Microcosm*

vii. *Transforming Principles*

viii. *Symmetry Lost and Symmetry Found*

ix. *Radical Uniformity in Macrocosm*

x. Quest

Introduction

From Pythagoras measuring harmonies on a lyre string to R. P. Feynman beating out salsa on his bongos, many a scientist has fallen in love with music. This love is not always rewarded with perfect mastery. Albert Einstein, an ardent amateur of the violin, provoked a more competent player to bellow at him, "Einstein, can't you count?"

Both music and scientific research, Einstein wrote, "are nourished by the same source of longing, and they complement one another in the release they offer." It seems to us, too, that the mysterious longing behind a scientist's search for meaning is the same that inspires creativity in music, art, or any other enterprise of the restless human spirit. And the release they offer is to inhabit, if only for a moment, some point of union between the lonely world of subjectivity and the shared universe of external reality.

Although scientists in love with music are commonplace, musicians who give their spare hours to science are less so. (For one notable exception, see chapter 4.) This is not surprising. The beauty of an unfamiliar art form can be hard to see, and the music beloved by scientists has rarely been "modern music." Benjamin Franklin, a contemporary of Handel and Mozart, was able to give some very scientific-sounding reasons why the sweet Scottish folk tunes he learned as a boy were the height of musical creation. Enjoyment comes with experience, so our grandchildren may be as perplexed by our failure to enjoy a computerized sequence of electronic screeches as we are by our grandparents' distaste for Debussy and Gershwin.

It's a shame, though, that the gradual diffusion of scientific knowledge has not made the beauties of science more familiar.

One of the inspirations for this book was the idea that practicing scientists could learn from the great composers about making beauty accessible. In particular, we were intrigued by a classically simple form of organization: the theme with variations. We have accordingly set out in the ten sections of this book to develop ten major themes of physics, showing through a series of variations how one simple idea can give birth to a logical, but delightfully surprising, series of interpretations.

Of course, each section begins with a very clear statement of its theme. But we also wanted to use the musical idea of a *prelude* whose harmonic excursions define the key before a fugue begins. Here the prelude to each section is a brief essay whose apparent digressions into anecdote or speculation are meant to set the stage for the developments to come. We have also interspersed the chapters with lighter *intermezzi,* diversions that are neither integral nor irrelevant to the subject at hand—just too intriguing to be altogether omitted.

The themes we have chosen vary, interweave, and build, much like the themes in a sustained musical work. For instance, the uniformity of nature's basic building blocks forms our first theme; the radical precision of this uniformity and the deep reason for it form our sixth theme. A parallel discussion of the uniformity of nature on the largest scales, and the radical meaning of this cosmic uniformity, gives rise to our second and ninth themes. And the symmetry of physical law, our eighth theme, abstracts natural uniformity from microcosm and macrocosm into the realm of thought.

You will find many such connections and resonances as you proceed through our book. For this reason, we urge you to be patient and to read through it whole, even if you're dying to learn about gluons, inflationary universes, familons, axions, and photinos. This book is not meant to be read for the latest dope on a few hot fields of research—although it contains much that is new, even to the point of being unfinished. The exciting new ideas of science are no less exciting, and much more meaningful, viewed in the context of the grand traditions they continue.

The traditions of science have been carried on over many centuries by the cooperation of thousands of ardent enthusiasts, including some of the most brilliant minds of every generation. In fact, science is the supreme example we have of a progressive human endeavor—like a medieval cathedral on a scale unbelievably grand. To enter this construction, even to peer through the doorway, is to be enriched.

We hesitate to join the earnest souls who treat learning about science (or, for that matter, music) as a somber duty of the cultured man. But it does seem to us shortsighted for anyone to ponder the great questions, unavoidable for a thinking person, about the ultimate nature of reality, or its transcendent meaning, without first finding out what physical reality is. And the insight into reality that modern physics affords is neither bland and commonsensical nor mystical and vague. It is at once surprising and precise, full of puzzling discords that beg to be resolved once more to harmony.

The idea of harmonies in nature captures a metatheme that recurs, in different guises, throughout our book. Science begins and ends in the physical world of sensation, but in seeking to understand this world and predict its behavior, science imagines other worlds, ruled by logic but inaccessible to perception. Only small glimmers of such worlds peep out around the corners of reality. But knowing that the physical world supports—no, demands—their extraordinary beauty, we return to it with a new feeling:

> And the end of all our exploring
> Will be to arrive where we started
> And know the place for the first time.

Acknowledgments

Among the godparents of this work, we would like especially to thank our agents John Brockman and Katinka Matson, and our editor Mary Cunnane, for discerning signs of promise in its awkward infancy.

Among many friends we especially wish to thank K. C. Cole, Dudley Shapere, and Sam Treiman for most helpful comments on early drafts of portions of the book. The work of Katie Nelson and Otto Sontag helped to make the final stages of preparation painless, and to smooth many rough edges.

Frank also wishes to express his deep gratitude to two notable physicists, Richard Feynman and David Gross, for the inspiration their works and personalities have given him over the years. And Betsy would like to thank Mary Jane Marchand, Sylvia Mendenhall, and Georges Guy for sharing with her their lively affection for language as a vehicle of thought.

Last but not least, we want to thank our parents, Frank John Wilczek, Mary Rose Cona Wilczek, Joseph Murray Devine, and Clothilde Elizabeth Reo Devine, for getting us started.

The lines from T. S. Eliot's poem "Little Gidding" are reprinted by permission of Harcourt Brace Jovanovich from *Four Quartets* (New York, 1943).

LEGO® is a registered trademark of INTERLEGO AG.

Two pages from the *Particle Properties Data Booklet* have been reproduced with the permission of the Particle Properties Data Group at SLAC, and with the assistance of Mrs. Virginia Pritchett.

The bubble-chamber photograph in Chapter 19 is reproduced with the permission of Brookhaven National Laboratories, and with the assistance of Dr. Nick Samios.

Many of the figures were drawn by Bonnie Bright.

Uniformity of Parts

The world around us seems to contain a bewildering diversity of materials. And until fairly recently the composition of truly distant objects—planets, the sun, stars—was held to be unknowable. We have learned, however, that light, always our main messenger from the external world, carries a much richer message than our unaided eyes perceive. Properly seen, the light from luminous bodies reveals that whether they are found on Earth or in the most distant galaxies, they are all made from a few common building blocks.

Prelude One

REPLY TO KEATS
(ON THE RAINBOW)

The rainbow bridge between earth and sky is an ancient and power-ful symbol. Norse gods climbed the rainbow from Earth to Asgard; to keep frost giants out, the band of red was filled with flames. The Greek goddess Iris, wearing an iridescent gown of water droplets, sped across the rainbow with messages from Olympus. Noah's God assured him that the rainbow would stand between them as a giant mnemonic: "And it shall come to pass, when I bring a cloud over the earth, that the bow shall be seen in the cloud: And I will remember my covenant, which is between me and you and every living creature of all flesh; and the waters shall no more become a flood to destroy all flesh."

The gods who bound earth to sky with rainbows have largely vanished. Does this change imply a reduction in the beauty of na-ture? John Keats certainly thought so:

> Do not all charms fly
> At the mere touch of cold philosophy?
> There was an awful rainbow once in heaven:
> We know her woof, her texture; she is given
> In the dull catalog of common things.
> Philosophy will clip an angel's wings,
> Conquer all mysteries by rule and line,
> Empty the haunted air, and gnomed mine—
> Unweave a rainbow. . . .

Nor is Keats alone in his feeling that "cold philosophy" can only devalue an ancient symbol of great power.

We no longer take the old stories, however charming, as explanations of the rainbow. But have we therefore "reduced" its meaning and symbolic power? We think not, and in what follows we hope to show that rich new symbolic meanings arise both from the history and from the content of the modern explanation.

René Descartes took the decisive first steps toward the scientific theory of the rainbow. If we were seeking to fix a birthday for modern science, we might well settle upon November 19, 1619, when Descartes took shelter from the cold by crawling into a large, old-fashioned stove.* He spent the day in privacy, warmth, and darkness, meditating. When he emerged, he had a revolutionary vision of man's place in nature. Reason and observation alone, he was convinced, could explain all natural phenomena—revealed truth was superfluous; ultimate mystery or capriciousness, unthinkable.

The opening of his book *Meteorology,* which contains his theory of the rainbow, harks back to this vision:

> It is our nature to have more admiration for the things above us than for those that are on our level, or below. . . . This leads me to hope that if I here explain the nature of clouds, in such a way that we will no longer have occasion to wonder at anything that can be seen of them, or anything that descends from them, we will easily believe that it is similarly possible to find the causes of everything that is most admirable above the earth.

Descartes' faith in his vision was amply confirmed by the success of this theory of the rainbow. By shining sunlight through a big globe filled with water, Descartes satisfied himself that spherical raindrops would separate one color from another in orderly fashion. (In any rainbow, or spectrum, we find the light broken down to red, orange, yellow, green, blue, and violet, always in this sequence. Like a prism, a water drop bends red light the least and violet the most.) Using simple experiments and calculations, Descartes found he could also predict just where in the sky the bow appears, and even why and where a secondary bow is sometimes visible. Descartes' vision of rational nature remains alive and vigorous today; the rest of this book is a testament to it.

Two centuries later Joseph von Fraunhofer (1787–1826), a contemporary of Keats, took the next decisive step in appreciating the message of rainbows.

---

*There is some scholarly dispute as to whether the term Descartes uses should be translated as "stove" or as "small warm room." We have opted for the more romantic version.

Fraunhofer discovered that the apparent continuity of a rainbow is an illusion. There are tiny gaps, dim or black arcs of missing colors, too narrow for us to see in the glare of natural rainbows. To say it another way, there are specific colors (specific wavelengths of light) in which sunlight is deficient. Fraunhofer eventually cataloged 576 of these gaps, or "absorption lines": 576 specific wavelengths missing from sunlight. Today tens of thousands are known. Unfortunately, Fraunhofer's career of discovery (like that of Keats) was cut short by consumption.

Similar gaps are found in the light from other stars, or from hot gases here on Earth, when their colors are spread out and analyzed. In fact, such artificial rainbows have patterns of relative brightness as characteristic of the substance emitting the light as a fingerprint is characteristic of a person. For example, two dark lines found in the violet region (at wavelengths 3.96×10^{-5} and 3.93×10^{-5} centimeters, to be exact), very prominent in the sun, are caused by the presence of calcium. By analyzing the light from a star, by studying the relative strength of different colors in the artificial rainbow it casts, we can learn what the star is made of, its temperature, and even such details as the strength of the magnetic field at its surface.

Most important, we learn from this analysis that star-stuff is the same as Earth-stuff. For example, every absorption line (and every pattern of such lines) ever found in the light of any star or nebula has been duplicated by a material in the laboratory; this precise correlation has never failed in tens of thousand of examples. On Fraunhofer's tomb, the epitaph reads *Approximavit sidera* (He brought closer the stars). It seems well deserved.

If scientific understanding of the rainbow has undermined some of its old charm, this understanding also suggests new and glorious symbolic meanings. The rainbow should symbolize for us that the world is comprehensible and that heaven and Earth are one.

1

The Nature of Color: We Are All Color-blind

I would observe that the important part of the theory is not that three elements enter into our sensation of color, but that there are only three. Optically, there are as many elements in the composition of a ray of light as there are different kinds of light in its spectrum; and, therefore, strictly speaking, its nature depends on an infinite number of variables.

—J. C. Maxwell, "Theory of the Perception of Colors"
(1856)

Science is based upon the evidence of our senses; and yet it transcends them. This paradox is well illustrated by the theory of color.

Science early discovered that light "in itself" has a much richer structure than our sense of vision reveals. It therefore becomes necessary to distinguish <u>physical color</u> from <u>sensory color</u>. It is possible to predict the sensation people will report when a given bundle of light rays impinges on their eyes—that is, the sensory color—in terms of the physical colors present in the rays. On the other hand, there are combinations of light that are definitely different but look the same. Physical color is, in this precise sense, more fundamental than sensory color, even though the latter is what we actually see.

The modern notion of physical color originates from a series of simple but elegant and profound experiments by Isaac Newton. In Newton's time it was already well known that a ray of white light—in practice, sunlight—emerges from a prism as a band of colors, a sort of artificial rainbow (figure 1.1). The leading explanation offered at the time was that light somehow became "degraded" by passing through the prism, that the glass in the prism absorbed parts of the white light, leaving colored rays behind. The idea behind this theory is not entirely silly; in fact, we know of many cases in which selective absorption does produce color effects. Stained glass works this way, as do most dyes. Nevertheless, Newton doubted that absorption is responsible for the colors produced when light passes through prisms—not least, we imagine, because prisms are made of the very paradigm of a material that does not absorb light: transparent glass.

Newton proposed an alternative explanation. He agreed that a beam of natural "white" light already contains a whole variety of colored light beams, but he had quite different ideas about what the prism does. The role of the prism, according to Newton, is merely to spread out preexisting colored beams, which overlap in the original white light.

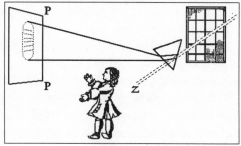

Figure 1.1

Figure 1.1 is based on an illustration to Voltaire's *Elémens de la philosophie de Newton,* published in 1738. The small figure of an eighteenth-century gentleman (trousers for men were still very far in the future) indicates a screen where the prism has separated out the colors in a narrow beam of sunlight to form a colored spectrum. The separation arises because beams of different colors are bent (or, to use the conventional term, refracted) through slightly different angles as they enter and leave the prism.

Newton designed simple experiments to test his ideas and, if they worked as he anticipated, demolish the old theory. Figure 1.2, based on an illustration in Newton's own great work *Opticks,* shows a stripped-down version of the crucial experiment.

A beam of white light is allowed to pass through not just one but two prisms, the second one reversed with respect to the first. The light emerges essentially as bright as when it started, and not col-

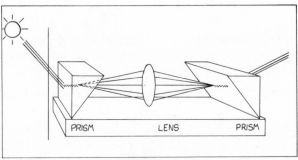

Figure 1.2

ored at all, just slightly displaced from its original path. The second prism reverses the deflections produced by the first. The different-color light rays that get separated by the first prism are recombined by the second into a beam of white light indistinguishable from the original beam. Contrary to the old ideas, nothing has been "subtracted" from the light in its passage through the two prisms.

If white light can be divided into components by passage through a prism, it is logical to ask next whether these components are fundamental. Beams of "pure" color can be created by putting a screen containing a small hole behind the prism, in such a way that only one color passes through the hole. Can such "pure" beams be broken down further—by passing them through other prisms, say, or perhaps by some other process? Newton tried many experiments of this kind and described his conclusions as follows:

> When any one sort of Rays hath been well parted from those of other kinds, it hath afterwards obstinately retained its colour, notwithstanding my utmost attempts to change it. I have refracted it with Prismes, and reflected it with Bodies, which in Day-light were of other colours; I have intercepted it with the coloured film of Air interceding two compressed plates of glass, transmitted it through coloured Mediums, and through Mediums irradiated with other sorts of Rays, and diversly terminated it; and yet never produced any new colour out of it.

In other words, the color of such pure beams is a fundamental, stable property.

The role of pure colors in the theory of light is similar to the role of elements in chemistry. As the chemical elements are the simple building blocks, themselves irreducible, out of which all other chemical substances are made, so the pure colors are the simple components from which all bundles of light are made.

The analogy between pure colors and chemical elements is very close. It is, however, worth mentioning two significant differences, one essential and one practical.

An essential difference is that whereas there are only a finite number of chemical elements, as far as we know there is a continuous infinity of possible pure colors. Some light is refracted through each of a continuum of slightly different angles emerging from the prism, and the light at each different angle represents a different pure color.

The practical difference is that it is much easier to analyze light into its elements than similarly to analyze chemical substances. Large books have been, and are still being, written on methods of chemical separation; the procedure can become very elaborate and specialized in some cases. For light, on the other hand, there is a simple, universal method of analysis—just pass it through a prism!

Now we can better appreciate the notion of physical color. The physical color of a bundle of light is defined by the amount of each of the pure or elementary colors it contains. It is like the chemical formula of a substance, which specifies how much of each chemical element the substance contains.

Having gained this insight into the fundamental physical nature of color, we should try to understand our sensation of it in the same language. The most important fact about human color vision is this: the complete range of sensory colors can be reproduced by suitably combining just three pure colors.

This fact, proved by experiments in the nineteenth century, now has a direct and appealing physiological explanation. In the human retina are three types of color-sensitive pigment molecules that change their shape when light impinges on them. The primary events in color vision are just these changes of shape. These changes trigger further chemical changes and ultimately lead to nerve impulses that somewhere further down the line, in the brain—by a process still very poorly understood—give rise to the sensation of color. Thus, in a meaningful sense, what we get to "see" is not light itself but its effects on these three types of molecules.

Now, each of the three pigments has its own characteristic sensitivity. One type is most likely to change shape in response to light with physical colors in the blue region, another in response to physical colors around yellow, the third in response to physical colors around red. In sensing what we do, namely, the number of pigment molecules of each type that change their shapes, we get information about three different averages of the intensity of the incoming physical colors. The human eye cannot extract any more information about color, beyond these three averages. In particular, combinations of incident physical colors that produce identical averages will give rise to the same sensation. Since just three different things about the incoming colors are sensed, it is reasonable—and can be proved mathematically—that by combining just three pure colors in varying intensities the whole range of possible sensations can be reproduced.

The ability of physically different bundles of light to produce identical sensations of color is exploited in color photography and color television. In each case, three ingredients—three dyes for photography, three types of excitable luminous molecules for television—are combined to give the full range of sensory colors.

Color-blind people lack one or more of the pigments. There are various identifiable types of color-blindness, depending on which pigments are missing. The full range of color sensations of a person missing one pigment can be obtained by combining at most two pure colors. A person missing two pigments—that is, having but one— sees the world in monochrome. Bundles of light that are indistinguishable to the color-blind can be distinguished by those with normal color vision, whose eyes extract more information from the incoming light.

In a larger sense, though, we are all color-blind. At best, we perceive three averages from an infinite manifold of physical colors. Why has nature been so parsimonious? Why don't we carry around little prisms in our heads, which would enable us to separate the colors and extract much more information? Or, more practically, why aren't we equipped with more pigments? Probably, it is because the pace of evolution of human intelligence—a striking anomaly to which we shall allude again in prelude 9—has outstripped the evolution of appropriate sensory systems. Lower animals have little use for the vast amounts of information improved color vision could provide. Indeed, a very substantial fraction of the human brain— perhaps 10 to 25 percent, depending on how you count—is devoted to processing visual information. Man's intelligence, which might enable him to use the extra information, has developed so suddenly, and so recently in evolutionary time, that the physical apparatus of his eyes has had no chance to catch up.

Fortunately, we need not wait for evolution. Using our intelligence, we can design instruments to relieve our sensory deficits—to improve our color vision. We can, so to speak, give ourselves better eyes. To those better eyes, which exploit the full potential of the information encoded in light, new worlds are revealed—read on.

2

Spectra: Music of the Spheres

*The heavenly motions are nothing but a continu-
ous song for several voices (perceived by the intel-
lect, not by the ear); a music which, through dis-
cordant tensions, through sincopes and cadenzas,
as it were (as men employ them in imitation of
those natural discords) progresses towards certain
pre-designed quasi six-voiced clausuras, and
thereby sets landmarks in the immeasurable flow
of time.*

—Johannes Kepler, *Harmonice mundi* (1619)

The *Harmonice mundi* of Johannes Kepler (1571–1630) is a milestone
in the history of science. In it, what is now known as Kepler's third
law of planetary motion is announced. This law, which relates the
speed with which a planet moves to its distance from the sun, provided the key to Newton's deduction of the law of gravity. More than
three hundred years later, Kepler's third law still appears at the
frontier of science—it is central to the modern "dark-matter" problem, as we shall see in chapter 31. If we actually turn to the *Harmonice mundi* itself, however, we find that Kepler's great discovery is
merely one small ingredient in a wild potpourri of number mysticism, music, and astrology, of which Arthur Koestler remarked,

> The *Harmony of the World* is . . . the climax of his lifelong obsession. What
> Kepler attempted here is, simply, to bare the ultimate secret of the uni-
> verse in an all-embracing synthesis of geometry, music, astrology, astron-
> omy and epistemology. It was the first attempt of this kind since Plato,
> and it is the last to our day.

Kepler was obsessed with an idea that goes back to Pythagoras
and the Orphic mystery religions of ancient Greece, and is a recur-
rent theme in Renaissance poetry and literature. This idea is that

the workings of the world are governed by relations of harmony and, in particular, that music is associated with the motion of the planets—the music of the spheres.

The notion of "music of the spheres" has historically been a vague, mystical, and elastic one; but we can start to define and appreciate it with the following thoughts. The bases of music are rhythm and harmony. Rhythm is ordered recurrence in time; we say two players are *in rhythm* if they are hitting notes at the same time or, more generally, if there is an orderly relation between the times when they play notes—a steady bass may underlie an intricate melody. As the planets move around the sun, they repeat their orbits periodically; thus there is already a primitive kind of rhythm in their motion. Kepler wished to find relations between the motions of the different planets, to show that in some sense they are all moving to a single cosmic rhythm.

Harmony, we now realize, can be considered a special form of rhythm. Sound is vibrations of air; pure musical tones are produced when the vibrations are of a particularly simple and regular form— that is, when they are periodic or, to say it another way, when they repeat themselves regularly in time. Two tones *harmonize* if their intervals of repetition are *in rhythm*—or, in more mathematical language, if their periods are *in proportion*. Kepler naturally did not have our modern understanding of the nature of sound, but in the third book of *Harmonice mundi* he attempted to make other, difficult to understand but perhaps somehow related, connections between musical harmony and mathematical proportion.

Kepler's obsession sustained him through years of tedious calculations and numerous failures. The successful form Kepler finally gave to his third law (the square of the time it takes for a planet to complete its orbit around the sun is proportional to the cube of its distance from the sun) is probably not the first relation anyone would guess exists between the orbits of different planets, and it was certainly not Kepler's. Some passages from Kepler's books, including the one that heads this chapter, read like tracts by "crackpot" amateur scientists or even outpourings from the delusional systems of schizophrenics. There is one all-important difference, though— Kepler worked incredibly hard to compare his conjectures with the accurate observations recorded by Tycho Brahe and was not satisfied until agreement was found. Small anomalies in the observed orbit of Mars led him to abandon centuries of precedent, and years of his own previous work, and propose that the planets move in ellipses rather than in perfect circles.

Is the vision of Pythagoras and the Orphics, Kepler's obsession, in any sense a real feature of the world, or is it only a historical curiosity? Is there truly a "music of the spheres"? We now argue that this marvelous dream is in fact closely realized in the physical world. The spheres, however, are not planets but electrons and atomic nuclei, and the music they emit is not in sound but in light.

Music in light—what does it mean? Let us recall our discussion of the nature of light in the preceding chapter. We (or rather Newton) found that the fundamental elements of light are what we call pure colors. These pure colors should be compared to pure tones in music. Arbitrary bundles of light are composed of mixtures of the pure colors, just as arbitrary musical chords can be composed of pure tones. The analogy touched on here, between light and sound as physical phenomena, runs extremely deep, as we shall have many occasions to observe in the following pages.

Our sensory systems, however, process light and sound quite differently. When we hear a musical chord, the individual tones that make up the chord do not lose their identities; each is still sensed. It is different for light: our eyes do not sense each individual basic pure color, but only three averages over them. In this regard, vision is much cruder than hearing. But by separating light into its component pure colors—for instance, with the help of a prism—and recording how much of each is present, we retain the information in light that our eyes ordinarily throw away. It's a truly wonderful thing, that it's possible to recover the lost information. It is as if we were all tone-deaf and had suddenly discovered a way to appreciate music.

So let us listen to the light—what music do we hear? For one thing, we can elicit from each chemical element its own, unique chord. You may sometime have noticed that a bright yellow flash is produced if ordinary table salt is sprinkled on a flame; in any case, this is a simple and effective experiment to try. This yellow flash is a first, bare hint of the huge subject of *flame spectra*. Sodium atoms in the salt, when excited by the heat of the flame, emit light that we sense as yellow. A more refined view, analyzing the light into pure colors, reveals something remarkable. If you pass the light from a sodium flash through a prism, you get a pattern very different from the familiar continuous rainbow that Newton elicited from natural sunlight. Instead of a continuous pattern, in which all gradations of pure color are apparently represented, the sodium flash generates a series of lines of light. In other words, only a certain discrete set of pure colors is emitted. Or, to put it yet another way, in the musical analogy, sodium produces a chord where sunlight produced all possible tones—"white noise." The yellow appearance of the sodium flame is caused by its particularly strong emission of a specific pure color in the yellow region (wavelength 5.89×10^{-5} centimeters, to be exact). This same emission is responsible for the yellowish tinge of sodium street lamps.

Other elements produce other chords. The fact that different elements emit light with different color characteristics is exploited by the makers of fireworks. Lithium produces a spectacular crimson flame; barium produces lime green. In each case, this is so because the color chord of the element has a very strong component in the corresponding region of the spectrum.

If our eyes were more perfect, we would see the atoms sing. A race of beings who had this sort of direct experience would no doubt include a high proportion of poets and atomic scientists.

3

Earth-stuff and Star-stuff

> [T]he mathematical thermology created by Fourier
> may tempt us to hope that . . . we may in time
> ascertain the mean temperature of the heavenly
> bodies: but I regard this order of facts as for ever
> excluded from our recognition. We can never learn
> their internal constitution. . . .
> —A. Comte, *Positive Philosophy* (1835)

Joseph von Fraunhofer became, despite his unpromising origins and early death, a remarkable figure in the history of science. He was apprenticed at twelve to a glass polisher who scoffed at the orphan boy's ambition to train himself to be a spectacles maker. Isolated, lonely, able to study his few books only at night, Fraunhofer at fourteen was rescued from drudgery by a freakish accident: his lodging house collapsed, trapping him in its wreckage. Helpers and sightseers, including the Elector of Bavaria, swarmed to the scene. The Elector was moved to present the grimy, disheveled boy, upon his final extrication, with a gift of eighteen ducats. Fraunhofer at once bought his way out of his apprenticeship, purchased books on optics and a glass-working machine, and began his brief but magnificent scientific career.

Fraunhofer did not, however, set himself the goal of pure scientific research. His aim throughout his career was technological: to perfect the design and construction of such optical instruments as telescopes and lenses. People often think of technology as the stepchild of basic science, but in reality each invigorates the other. It was Fraunhofer's mastery of the technology of optics, his superb craftsmanship, that enabled him to make one of the most significant discoveries in the history of science.

We have already referred to his seminal discovery that there are gaps in the spectrum of the sun. In other words, the apparent continuity of the rainbow turns out, on sufficiently close inspection, to be an illusion. Colors are missing; there are minuscule arcs of black embedded in the rainbow. This discovery more than any other allowed astronomers to do what Comte asserted would never be possible—to determine what stars are made of.

To understand what Fraunhofer's discovery means, and how it has been extended in many directions since, we pause to consider further the physics of spectral lines.

Fraunhofer's missing colors suggest a sort of negative image of the flame spectra. In fact, Fraunhofer noticed that the position of many of his gaps corresponded precisely to the colors emitted in flame spectra. He suspected a causal connection, but died before completing this research.

What is the connection between Fraunhofer's pattern of "holes in the rainbow" and the bright lines of flame spectra? We now know that they are dual manifestations of the same fundamental process, the exchange of energy between light and matter. When a substance is heated and radiates light into empty space, we see a flame, or emission, spectrum. When the same substance is kept cold, it tends to absorb exactly the same pure colors it radiates when hot. In each case, energy flows between light and matter, from the hotter to the colder stuff. The bright lines of emission and the dark lines of absorption both indicate that certain colors of light exchange energy with a given substance more readily than do others.

Consider a soprano who wants to break a crystal goblet using her voice. Her first step is to strike it, make it ring, and by listening learn what tone it particularly likes to emit. She knows it will absorb sound of that same pitch most efficiently. So she tries to produce the exact same tone she has heard; in this way, she can efficiently pump energy into vibrations of the goblet and shatter it. The soprano is implicitly making use of the same connection between emission and absorption for sound that we have just discussed for light. She has taken an emission spectrum and relied on the fact that absorption will be most efficient for the same tones.

Just as the resonant tone of a goblet is determined by its shape and composition, the colors resonant with a bit of matter—the ones that appear in its spectrum—are determined by the internal shape and dynamics of its constituent molecules. The emission or absorption of a sample of sodium (for example) will always contain a predictable set of lines utterly distinct from those of any other substance.

It became a great enterprise in the nineteenth century to determine the emission spectra of the elements, and also of molecules. The resulting catalog of spectra shows its worth when you wish to determine the composition of some otherwise unknown material. By studying the relative intensity of the colors in the light it emits, you learn—comparing with the catalog—what elements and chemical molecules it contains. Each element, each molecule that is present sounds its own, characteristic chord.

We can apply the same principle to light from the sun, or to light from other stars. The hundreds of features in any star's spectrum contain a wealth of information, patterns that can be compared to the spectra produced by known substances in the laboratory.

Many interesting discoveries have been made by following this program. But perhaps the most profound conclusion is simply this: that the sun and stars are made out of the same stuff as we find here on Earth. By now, not only Fraunhofer's 576 lines but also thousands of weaker ones in the solar spectrum have been matched with corresponding features in the spectra of earthly materials. The same sort of matching has been done, in less detail, for many other stars and for other celestial objects.

A few examples will illustrate how fruitful has been the process of showing, in honest detail, that the materials making up the cosmos are exactly the same as the materials we find on Earth.

Helium takes its name from the Greek *helios,* for "sun," because it was actually observed in the sun before it was discovered on Earth. Several bright yellow emission lines in the sun's spectrum were first detected during the solar eclipse of August 18, 1868. These lines from the hot but tenuous chromosphere around the sun are normally lost in the overall glare, but they become easily visible during eclipses. It was quickly recognized that these particular lines did not match the emissions of any known terrestrial substance. It was not until 1895 that helium was found on Earth.

Helium is hard to detect on Earth because it is both rare and chemically inert. However, later research has shown that Earth is exceptional in this regard. Hydrogen and helium together constitute close to 99 percent of the mass both of the sun and of the universe as a whole.

Helium has come a long way from its cosmic origins to its popular use in inflating balloons. More significantly, it is helium, with its unique property of remaining liquid down to absolute zero, that makes modern low-temperature technology possible; ironically, this substance, first seen in the hottest visible parts of the sun, is now used on Earth as the ultimate refrigerant.

The radioactive element *technetium* does not occur naturally on Earth, because it is unstable. Its atoms spontaneously decay with a half-life of two hundred thousand years—a very short time compared with the age of Earth. In 1937, man-made technetium was produced at a cyclotron. Hence the name *technetium,* from the Greek *technetos,* meaning "artificial." In 1952, P. W. Merrill announced his discovery that the spectra of certain stars contain the characteristic lines of technetium. This beautiful discovery provides remarkably direct evidence that element building is taking place in these stars, because any "primordial" technetium would long since have decayed away.

Certain bright lines in the spectra of nebulae—bright clouds of gas heated to high temperature by the light of nearby energetic stars—resisted identification with the spectral features of any material known on earth for many years. A new element, *nebulium,* found only in nebulae, was held responsible, doubtless with the previous example of helium in mind. The situation became embarrassing, however, as increased knowledge of atomic physics made it clear that there was simply no room for new elements. Finally, in 1927, it was found that the lines in question were caused by atoms of oxygen that had been doubly ionized—that is, stripped of two of their sixteen electrons. Under normal (earthly) conditions, such ionized atoms quickly find electrons and revert to their electrically neutral form. Only in extremely hot and tenuous gases, gases so tenuous they would normally be considered "good vacuum" here on Earth, can the ions survive sufficiently long to be observable. Such, we learn, are the nebulae.

In every case, it has been possible to carry through a complete and detailed matching between cosmic and terrestial spectra. No room remains for doubt that star-stuff and Earth-stuff are the same stuff.

Rhapsody on $\mathcal{N}$ (The World Between)

The disproportion between human and atomic scales is enormous. Choosing a number to describe it depends on your choice of objects to compare, but the measure most commonly used is a number called Avogadro's number and denoted $\mathcal{N}$. It is the number of atoms in one gram of hydrogen, about 6×10^{23} atoms. More roughly, a gram of any substance contains about this many atoms.

Six followed by twenty-three zeros—it is deceptively easy to write down, but difficult to comprehend. It carries the implication that there can be worlds within worlds within what we think of as tiny amounts of matter. It is a number to conjure with.

For example, the average human brain contains something like 10^{12} cells. Some of these cells can do moderately sophisticated processing on the signals they receive. Still, surely we can imagine cleverly assembling 10^6—a million—atoms to make a machine that duplicates the function of one brain cell. (Think what a complicated Tinkertoy you could make with this many parts, even though each part is incredibly simple.) Let's hook 10^{12} of these "cells" together. We have now constructed, in our imagination, a human brain-equivalent computer weighing a small fraction of one gram. One pound will match the thinking power of the entire U.S. population. Since signals travel faster through smaller units, these tiny artificial brains would actually be more quick-witted than their natural counterparts, for equal complexity of connections.

Miniaturization of computers has come a long way recently, but you see that, because $\mathcal{N}$ is so big, there's still plenty of room to shrink.

More generally, it is the huge size of $\mathscr{N}$ that allows units of modest mass to combine the dual properties of stability and complexity. Stable subunits must contain a large number of atoms, for one thing so that they can work despite random molecular jiggling (Brownian motion) and, in general, so that the failure of a single component does not disable the whole. On the other hand, it takes many stable subunits to arrive at a useful complexity. Because $\mathscr{N}$ is so large, both requirements can be met simultaneously in small masses of matter. Perhaps the ultimate example: the information necessary reliably to assemble a human being fits on a DNA molecule weighing less than the ink in the period at the end of this sentence.

The mass of a typical star is about $\mathscr{N}$ times the mass of a human being (more precisely, about a thousand times this; here we are speaking with "astrophysical accuracy"), and there are about $\mathscr{N}$ stars in the accessible universe. So we pass from atomic scales to human with one factor of $\mathscr{N}$; multiplying twice more by $\mathscr{N}$, we encompass the accessible universe.

Does finding this many stars indicate a deep connection between microcosmos and macrocosmos? We shall find many deep connections between these opposite poles of existence, but this bit of numerical serendipity is not among them. There is nothing fundamental about the gram as a unit of mass; also, the number of stars accessible to vision changes with time. The equality is an incongruous coincidence, a cosmic jest, which nevertheless can serve as a useful orienting metaphor. If we regard stars as our units—our atoms—and are content to look at them from a distance, without inquiring about their internal structure, the universe becomes about as complicated to analyze as a gram of matter.

Fundamental physics often involves us in considering things smaller even than atoms or much larger than everyday objects. Physics has come to understand "what there is" in terms of building blocks smaller than atoms, existing in proportions and arrangements ultimately determined by processes on gigantic cosmic scales. Similarly, the fundamental rules governing "how things work" are rules for how the subatomic building blocks interact with each other; the behaviors of larger objects are found to be (as far as we can tell) consequences of these.

But while it is in some sense philosophically correct that the inter-

mediate scales present nothing fundamentally new—their phenomena are implicit in the lawful behavior of a much smaller microcosm, whose content is ultimately dictated by the history of a
much larger macrocosm—it is profoundly wrong from a human perspective. Finding the fundamental laws can be compared to discovering all the possible <u>notes</u> a piano can play. If some fellow played each
note for you once and then said, "Now I've shown you all there is to
music; the rest is just different combinations"—well, you'd be entitled to think him a joker or a lunatic.

One doesn't study quarks and cosmology in hopes of learning how
to make a good soufflé or how to interpret a friend's smile. Appropriate concepts of a different order are introduced to describe these
things, concepts that make little or no use of physics. Physics cannot
replace these concepts, but it can enrich them by putting them in the
broader context of its world picture. Perhaps, for example, we can
better admire the delicacy of the soufflé when we realize that its
production takes clever advantage of the normally random, agitated
motions of protein molecules as we heat them. Our friend's smile is
in no way diminished by the realization that it is caused by electrons
and ions moving in her brain; a better and more appropriate response is to marvel and rejoice at the wealth of meaning and structure that can emerge from arrangements of vast numbers of utterly
simple stereotyped units.

Uniformity of Structure

Although the universe contains a wonderful diversity of strange and peculiar objects, the uniformity of its content is more wonderful still. In the most remote parts of the universe, in every direction in the sky, we find matter organized into broadly similar galaxies of broadly similar stars. And these similar units are roughly uniformly distributed. Any big chunk of the observable universe looks pretty much the same as any other.

In looking far away, we also look backward in time, because the speed of light is finite. Messenger light—now transformed into microwave radiation—from ten billion years ago reveals that the universe has evolved from a hotter, denser, and even <u>more</u> uniform past.

We are challenged to understand what made the early universe almost perfectly uniform, and how from such beginnings it acquired the rich texture it has today.

Prelude Two

RADICAL CONSERVATISM

John Wheeler's career in physics has taken him in two widely diverging directions. He began as a theorist of thermonuclear explosions, then went on to greater fame with his theories of massive implosions of stellar matter, gravitational maelstroms for which he coined the name *black holes*.

"Radical conservatism" is Wheeler's semiparadoxical description of a style of thought that has many times proved its worth in physics.* The style has two components.

It is <u>conservative</u> in its reluctance to introduce new assumptions. This attitude was given its classic formulation by William of Occam, a medieval philosopher, and is known as Occam's razor. Occam's epigram "It is vain to do with more what can be done with fewer" became more famous in a fancier form: "One should not multiply essences without necessity."

Isaac Newton expressed himself similarly in a famous methodological passage:

> For anything which is not deduced from phenomena ought to be called a hypothesis, and hypotheses of this kind, whether metaphysical or physical, whether of occult qualities or mechanical, have no place in experimental philosophy.

This conservative, critical attitude by itself is rather bloodless. Creative tension and power is added to it by a <u>radical</u> approach to the few assumptions that are adopted. These assumptions must be

*The scientific style of radical conservatism, of course, has no necessary connection with the political style of the same name.

formulated precisely and pushed as hard as possible. Their consequences must be fully drawn; they must be applied to as many
situations as possible in the natural world and to the oddest and
extremest conditions we can set up in the laboratory. Here is Newton again, contrasting his mathematically precise law of gravity and
its use with the wordplay typical of Aristotelian physics:

> To tell us that every species of things is endowed with an occult specific
> quality by which it acts and produces manifest effects, is to tell us nothing:
> but to derive two or three general principles of motion from phenomena,
> and afterwards to tell us how the properties and actions of all corporeal
> things follow from those manifest principles, would be a very great step
> in philosophy.

Perhaps the greatest strength of radical conservatism is that it
occasionally leads to really surprising, yet inherently credible, predictions—which vague or merely plausible thinking never does.

Consider, as an example of the radically conservative approach, the
discovery of the planet Neptune. In line with the program of press-
ing assumptions as hard as possible, theoretical astronomers shar-
pened their mathematical tools over decades and calculated, on the
basis of Newton's law of gravitation, extremely accurate orbits for
the planets. A small discrepancy was found between the calculated
orbit of Uranus and the observed one. All sorts of hypotheses could
be imagined—perhaps the law of gravity was slightly different from
Newton's at large distances, or was not universal after all (Uranus
might be different . . .). The radically conservative procedure, however, is to stay with the existing law until different assumptions are
absolutely necessary. Two mathematical astronomers, John Adams
and Urbain LeVerrier, working independently, realized in the 1840s
that Newton's law could be maintained if there was a large planet,
hitherto unknown, in the right place. This planet was duly looked
for, and found, in the right place; it is now known as Neptune. Both
components of the radically conservative approach—pushing the
framework as hard as possible and yet not giving it up until faced
with a clear contradiction—were crucial in making this magnificent
discovery possible.

The early work of Einstein offers multiple examples. In 1905, he
made three major theoretical discoveries, in widely different areas

of physics. One thread that ties them together is their radically conservative style.

Einstein's first paper on the special theory of relativity is probably the purest illustration of this style in the whole literature of physics. He begins with, and proceeds to push very hard, two simple assumptions, taken from a large body of observations: that the speed of light in empty space is a constant of nature and that the laws of physics (in particular, the value of the speed of light) appear the same to all observers moving with constant velocity. These two assumptions seem, at first sight, contradictory. It is hard to imagine how a light beam can appear to be moving equally fast both to someone standing still and to someone "chasing" it. But by analyzing exactly what it means, operationally, to measure the velocity of light, Einstein showed that his two assumptions could both be maintained—at the price of upsetting some of our intuitive notions about space, time, and the addition of velocities.

Einstein's work on Brownian motion also stems from pushing two simple assumptions, both commonly used in the physics of his time, extremely hard. One of these is the idea that heat is just a particular form of energy, hidden motion on very small scales; the other is the existence of atoms. Einstein reasoned that the motion associated with heat would be a large and ceaseless agitation on scales close to atomic; he calculated the form of this motion and suggested how it could be observed. In reality, the ceaseless, random jiggling of pollen grains viewed through a microscope had been described much earlier by the botanist Robert Brown, but not properly interpreted.

Einstein's third discovery, the photoelectric effect, illustrates another advantage of radical conservatism—its power to self-destruct in a fruitful way. Einstein realized—after considerable struggle, we imagine—that certain experiments could in no way be reconciled with established theories of the nature of light. This realization, which only a radically conservative analysis could lead to, forced him to look for a fundamentally different way of thinking about radiation. He took the quantum ideas, previously advanced hesitantly and with some confusion by Max Planck, as fundamental assumptions to be be applied not only to radiation theory but to all of physics, in the best radically conservative style. Thus began his long, stormy, and fruitful love-hate relationship with quantum mechanics—of which much more below.

We can get another perspective on the radically conservative style by contrasting it with its opposite. Any fairly well known physicist gets letters from well-intentioned but not necessarily well-trained people proposing new theories of the universe or large parts thereof. Their theories usually make only passing reference to the body of assumptions painfully tested and refined in over three hundred years of modern physics. Instead, they propose wholly new, vague, and sweeping assumptions. "Can you disprove this?" they ask. Well, probably, if the ideas are stated precisely enough. But this isn't the right question, whether asked in this context or (for example) in that of parapsychology. A better question is "What would Occam make of this?"

Another, witty variation on this theme was invented by Wolfgang Pauli, a famous physicist and notorious wiseacre. A long discussion with physicist X left Pauli visibly unimpressed with X's ideas. Finally, X was moved to plead, "But surely, Pauli, you don't think what I've said is completely wrong?" to which Pauli replied, "No, I think what you said is not even wrong."

4

The Cosmic Order of Galaxies

It is not from space that I must seek to obtain my
dignity, but from the control of my thought. . . .
The spaces of the universe enfold me and swallow
me up like a speck; but I, by the power of thought,
may comprehend the universe.

—Blaise Pascal, *Pensées* (1670)

The scale of the known universe has been vastly stretched since Pascal's time. We have measured its empty spaces and counted its starry inhabitants, and have in each case found magnitudes that dwarf anything in our ordinary experience. Yet, even as scientific cosmology continues to expand the spaces of the universe, Pascal's bold leap of faith seems more appropriate than ever. To look at the structure of the universe on larger and larger scales is to see more and more clearly its deep simplicity and order.

This modern view of structural order in the universe rises from firm foundations. The story of its construction begins, paradoxically, with an obvious <u>nonuniformity</u> among the visible stars.

The Milky Way is clearly visible to the naked eye* as a band of light stretching across the night sky; it has been known since prehistory and figures in many mythologies. The true nature of this band of light was announced by Galileo, the first to use a telescope for astronomy, in his *Starry Messenger* (1610): "[T]he Milky Way is nothing else but a mass of innumerable stars planted together in clusters. Upon whatever part of it you direct the telescope, straightaway a vast crowd of stars presents itself to view. . . ."

*That is, in an environment free of air pollution and the glare of artificial light. Frank, who grew up in Queens, first saw the Milky Way when he was a teenager on a visit to the Rocky Mountains of Colorado. Standing beneath the night sky as it must have appeared to our distant ancestors, he remembers feeling that he had entered another state of consciousness.

Our own sun is a member of this great star system, or *galaxy* (to give the scientific term for an assemblage of many millions of stars linked together by gravity). For hundreds of years, however, it remained uncertain whether the Milky Way constitutes a complete and unique "island universe" or whether, instead, there are objects outside it. Immanuel Kant had already, in 1755, hit on essentially the right idea:

> I come now to another part of my system, and because it suggests a lofty idea of the plan of creation, it appears to me as the most seductive. The sequence of ideas that led us to it is very simple and natural. . . . [L]et us imagine a system of stars gathered together in a common plane, like those of the Milky Way, but situated so far away from us that even with the telescope we cannot distinguish the stars composing it. . . . [S]uch a stellar world will appear to the observer, who contemplates it at so enormous a distance, only as a little spot feebly illumined and subtending a very small angle. . . . The faintness of its light, its form, and its appreciable diameter will obviously distinguish it from the isolated stars around it.
>
> We do not need to seek far in the observations of astronomers to meet with such phenomena. They have been seen by various observers. . . . Their analogy with our own system of stars; their form, which is precisely what it should be according to our theory; the faintness of their light, which denotes an infinite distance; all are in admirable accord and lead us to consider these elliptical spots as systems of the same order as our own. . . .

These inspired speculations of Kant were not, however, backed up by any hard evidence.

The fundamental problem is to determine the distance of astronomical objects. The attempts of William Herschel (1738–1822) to map the skies illuminate the difficulty involved. Herschel is an extraordinary figure on many counts. He was a talented musician and composer whose fascination with musical harmony led him to fall in love with optics and astronomy. For many years, he supported himself, and his sister (and fellow astronomer) Caroline, with his earnings as organist at a fashionable cathedral in the English resort of Bath. By night, they watched the skies or ground enormous mirrors for new telescopes. Herschel's discovery of the planet Uranus won him enormous fame, and modest fortune. His greatest achievement, however, sprang from his ambition to survey the universe <u>beyond</u> the solar system.

Herschel wanted to find the shape, in three dimensions, of the star system around us. To do this, he needed some way of estimating the distance to each star.

The powerful technique he devised is still in use today, and still called by the old-fashioned name of *standard candles.* The basic idea is simple. Suppose all (standard) candles give off the same amount of light. Then, if you want to know how far away some candle is, you can figure out its distance by measuring its <u>apparent</u> brightness, and comparing this with the apparent brightness of a similar candle whose distance you know.

In radically conservative style, Herschel made the bold assumption that all stars have the same intrinsic brightness. The "standard candle" Herschel used for his measurements was Sirius, the dog star. (Sirius is the brightest star in the sky, and one of the few stars close enough to Earth that its distance can be estimated by *parallax.* That is, its position relative to more distant stars appears to shift as Earth's motion around the sun lets us view it from opposite ends of our orbit.) To estimate the distance to some other star—say, Altair— Herschel needed two telescopes, as precisely similar as he could make them, and a bit of ingenuity. Pointing one telescope at Altair, he would then fix the second on Sirius and keep on draping sheets of muslin over its end, dimming Sirius step by step until both images looked equally bright. In essence, he measured stellar distances by counting sheets of muslin.

Modern astronomers have enormously powerful telescopes that focus tiny beams of starlight into sensitive measuring devices, or record them forever on photographic plates. To turn their stacks of numbers and pictures into stellar distances, however, they still use the standard candle technique pioneered by Herschel. (With one important refinement—today we compare only stars that have similar spectra; such similar stars do have roughly equal intrinsic brightness.)

After centuries of star mapping, the Milky Way stands revealed as an object of great beauty. It combines a variety of lovely shapes. A rosy sphere of ancient stars lies at the center of a vast, flattened disk of gas and stars in many colors, a disk that includes our own small, yellow sun. Also within the disk, two glowing spiral arms of cosmic dust, where bright new stars are being born, lie twined around the center.

Although we now know it is not the whole universe, the awesome image of our Milky Way as a glittering stellar island in a sea of black remains.

Even fellow members of our Milky Way are so far away that the very closest star appears as no more than a point of light against night's blackness. If you are a stargazer, however, you may have noticed that between Cassiopeia's *W* and the great square of Pegasus lies a small, hazy patch of brightness. This patch is the light from the great Andromeda galaxy, more than two million light years away—the most distant object visible to the naked eye. But we're getting ahead of our story.

Many such patches of light have been revealed by telescopes, and these were what Kant guessed might be other island universes. But how can you tell whether the smudge you see through a telescope is the accumulated light of billions of stars, made to seem faint by immense distance, or instead a relatively modest source of light close by?

Using stars within our galaxy as standard candles to compare with blobs of light of unknown nature seemed problematical at best. The problem was finally solved thanks to a special class of bright variable stars that proclaim their identity even across intergalactic distances. These are the *Cepheid variables*. Cepheids, as they are called, are found sprinkled about in galaxies throughout the universe. Their brightness waxes and wanes, in cycles (periods) of from two to forty days, depending on the particular star.

From studies of nearby Cepheids, whose distance can be set equal to the known distance of neighboring normal stars, a most significant regularity emerges: Cepheids with the same period of variability have the same intrinsic brightness. Furthermore, the intrinsic brightnesses of Cepheids with different periods are related in a simple way—the longer the period, the brighter the star. Because the Cepheids as a class are unusually bright stars, and because they are variable, they are readily identified wherever they occur, out to very large distances. The period of variability, which is readily observable, tells us the intrinsic brightness, so Cepheids make ideal standard candles.

With the help of the 100-inch telescope on Mount Wilson, Edwin Hubble (1889–1952) was able to identify Cepheid variables in the Andromeda "patch of light" and in many other objects we now know to be external galaxies. (Research in those days moved at a leisurely pace. Hubble had gathered his data by 1923, announced his results at Oxford in 1925, and finally published them in 1929.) By establishing the distance to these astronomical objects, Hubble proved that they had to be far outside the Milky Way galaxy. For galaxies so far away to appear as bright as they do, they must be intrinsically very bright indeed. They are great star systems themselves, comparable to our own Milky Way.

And so Kant's inspired speculation was proved true, nearly two centuries after he had made it.

Following this breakthrough, it became clear that the common unit of cosmic organization is the galaxy. Galaxies are not all identical, but they come in a limited range of sizes and shapes. Although not clones, they are clearly members of the same family.

Between the galaxies lie vast intergalactic spaces that emit very little light and exert very little gravitational pull. We are therefore tempted to say that the intergalactic spaces are empty. Although we shall later have occasion to qualify this statement (chapter 31), it remains true that the visible universe is organized rather neatly into galaxies.

The distribution of galaxies is irregular in detail. They can be found singly, in small groups, or, occasionally, in giant clusters. When very large regions of the sky are compared, however, these irregularities and clusterings average out. Any sufficiently large region is very much like any other. No new forms of organization appear; there are no distinguishing structures. This uniformity contrasts dramatically with the distribution of observable stars, which are concentrated in the band of the Milky Way. In short, the farther we have looked out into the enfolding spaces of the universe, the more we have found it possible to comprehend it.

Galactic uniformity is both a gift and a riddle. A gift, because it means that any large chunk of the universe is a "fair sample"; the remainder just repeats it. Understand one chunk, and you understand the whole thing. A riddle, because why does the formation of structure stop and nature become content to repeat herself? Researchers have been attempting to exploit the gift, and wrestling with the riddle, ever since—not without success, as later pages testify.

Doppler Shift

Sometimes a few lines in a scientific biography will conjure up a brightly colored human moment—Archimedes, leaping naked from his bath to shout his triumph through the streets of Syracuse—Joseph von Fraunhofer, receiving the gold that will buy his freedom from a hated master—William and Catherine Herschel in their observatory late at night, brother and sister gazing out toward infinity through separate telescopes. We have tried in vain to discover such a moment in the obscure and frustrated life of the Austrian physicist Christian Doppler (1803–1853). The effect he discovered has made his name a byword among astronomers today, but his contemporaries derided him for the errors in its formulation. Doppler eked out a living for his family by teaching mathematics in secondary schools and a mining college. In 1850, he finally attained a professorship at Vienna. Three years later, he was dead of consumption.

The effect Doppler discovered, most commonly known as the *Doppler shift,* is, quite simply, a kind of cosmic speedometer. By measuring the Doppler shift of a telescopic image, we can tell just how fast the object we see is moving away from us. To Doppler's contemporaries, it seemed a minor scientific curiosity; to ours, it has proved an essential tool for mapping out the universe.

To bring out the underlying simplicity of this effect, we first consider a situation that might seem remote from physics or cosmology but will turn out to be very much to the point. Imagine an army on campaign, sending messages to its king, who has stayed at home. Messengers depart every hour and ride directly without stopping to the king. Now, if the army is advancing—that is, moving away from the king—each messenger will have a slightly longer distance to cover than the preceding one. As a result, the king receives these messages more than one hour apart. Even though the messages are being sent every hour, the interval is "stretched out" because increasing times are needed for the horsemen to cover ever-longer distances.

Conversely, if the army is in retreat—being driven back toward its home base and its king—the king will receive messages more frequently than once per hour, for similar reasons. In either case, of course, the size of the effect depends on how fast the army is moving. The messages from a rapidly advancing army will be considerably stretched out, and the messages from a rapidly retreating army will be squeezed together. In the extreme case where the army is in wholesale retreat, moving back as fast as the messengers ride, successive messengers will obviously arrive together. Good news dribbles in but bad news comes all in a rush.

Without even reading the messages, therefore, the king can tell something about the way his army is moving. The intervals between the messages serve to tell whether the army is advancing or retreating, and how fast.

(It should be noticed that these intervals give information only about radial motion—that is, about motion toward or away from the king. If the army is circling the capital, successive messengers will take different directions toward the king, but they still cover the same distance and their messages arrive spaced by one hour. So, speaking more precisely, what the king can extract from the intervals between messages is not the total speed but the radial component of velocity—or, more plainly, the rate at which the army is moving toward or away from him.)

So much for kings and armies. The arguments are general. They apply without change whenever regular, periodic messages are sent at finite speed from a moving source. If the source is moving away from us, the period of the message we observe is longer than the period at the source; conversely the message from an approaching source has its apparent period decreased. It is usual and convenient to speak of the *frequency,* which is the number of events occurring in some fixed time. The term *frequency* is appropriate, because it tells you how frequently messages are received. We can rephrase our conclusions in this way: the frequency from a receding source is decreased; the frequency from an approaching source is increased.

Now let's turn to the corresponding effects for sound. Physically, a musical tone is a periodic vibration of air; the density and pressure of the air are made to vary in a regular way. "High" tones are, in this language, high frequency vibrations and "low" tones are low frequency vibrations. In terms of our metaphor, we can think of a musical tone as a particularly simple sort of message: we are being told how frequently the density of air reaches a peak. Messengers are dispatched once each period and "ride" at the speed of sound.

This leads us to anticipate the following effect: the musical tones we hear from a moving source, be it a tuning fork, an ambulance siren, or any other source of sound, will be different from the tone produced by the same source at rest. The tones we hear will be higher if the source is approaching us and lower if it is receding. The Doppler shift for sound was first demonstrated in 1845. A locomotive (then the emblem of modernity) was engaged to pull through Utrecht an open car, with several trumpeters all blasting out a single note. Observers heard a sudden drop in tone as the car passed by—a dramatic moment. Doppler probably wasn't even there; indeed, the demonstration, arranged by one of his severest critics, had most likely been intended to discredit his predictions.

Of most interest for us is the corresponding effect for light (also called the Doppler shift—is posthumous fame a consolation prize?). We have identified pure colors as the analogues, for light, of pure musical tones. Experiments show that the pattern of colors emitted by a source of light—its spectrum—is systematically altered by motion. If the source is moving away from you, the pattern is shifted toward the red end of the rainbow; if the source is moving toward you, the pattern is shifted in the opposite direction, toward the blue.

Light is therefore behaving as if it, like sound, is some sort of periodic vibration, transmitted at a finite speed. If the colors toward red are associated with lower frequencies, and those toward blue with higher frequencies, the optical shift can be explained in exactly the same way as the acoustic shift. As we go on, it will become apparent that this behavior is no mere coincidence; light is in fact a periodic disturbance propagated at finite speed, and the different pure colors are vibrations with different frequencies. The Doppler shift for light hints at its physical nature.

The Doppler effect for light is hard to observe in everyday life, because we rarely encounter sources whose velocities are anywhere near the speed of light. Because of this, the shifts are very small. However, with sophisticated modern instruments for analyzing spectra, even minute shifts can be observed. The radar guns policemen use to catch speeders take advantage of the Doppler shift. (*Radar* takes its name from radio waves, which are a form of invisible "light" at frequencies well below those that human eyes respond to.) By bouncing a radar wave off a moving car and then comparing the frequency of the waves emitted by the device to the frequency of the waves reflected back from the target, one can infer the speed of the target.

Astronomers use the Doppler shift for loftier purposes. If the whole pattern of spectral lines in the light of a star or galaxy is shifted to lower (or higher) frequencies, measuring this shift will tell us how fast the star or galaxy is moving away (or approaching).

In the next chapter, we will discuss the most dramatic use of the cosmic speedometer so far, to reveal that the universe is expanding. But before turning to this grand cosmic application, let us pause to complete an earlier circle of ideas.

Recall how Newton, in baroque prose, cataloged a variety of indignities to which he subjected light rays of a pure color without being able to alter their color. Now we have found that there is a way it can be done—namely, by moving. The Doppler shift is nothing but a change in pure colors, produced by motion. A pure blue ray, for instance, will appear as a pure red ray to an observer moving rapidly away from its source. (Newton could not observe this effect with the crude instruments available to him.)

It is very satisfying, philosophically, that different pure colors can be transformed into one another this way. It means that they are all intimately related to one another—each is in a strong sense equivalent to any of the others, viewed by a suitably moving observer. We therefore do not require an infinity of different physical entities corresponding to the different pure colors; we come to realize that they are all manifestations of just one underlying reality.

5

Expansion and Uniformity

Plus ça change, plus c'est la même chose.
—A. Karr, *Les guêpes* (1849)

Once Edwin Hubble had established that the galaxies are giant star systems similar to and well outside our own Milky Way galaxy, he immediately began a detailed study of their distribution in space and their physical properties.

We have already mentioned one important finding to emerge from this study: the uniformity of galaxies. The galaxies all bear a family resemblance to one another, and their distribution in space is, averaged over large regions, roughly uniform.

Hubble soon made a second discovery, just as significant and much more surprising. He found that the spectral patterns of distant galaxies tend to be shifted toward the red, low-frequency end of the spectrum, by an amount that grows with their distance. This systematic effect is just as would be expected from the Doppler shift, for galaxies moving away from us.

When the observations are treated quantitatively, a simple regularity known as Hubble's law emerges: distant galaxies are receding from us, with velocities proportional to their distance. Thus, if galaxy A is twice as far away as galaxy B, then, according to Hubble's law, galaxy A is also moving away from us twice as fast as galaxy B. All distant galaxies are moving away; their light is always shifted toward the red, never the blue. It is as if the scale of the universe, space itself, expands.

The phrase "expansion of the universe" is often used loosely. You should realize, however, that it does not mean that everything in the universe is expanding. It's not even clear what this latter idea could mean, since any "rulers" we use to establish distances would likewise expand and would therefore still give the same measurements. The precise sense in which the universe can be said to expand is slightly more subtle. The galaxies are like the pieces hurled from the exploding Death Star, which get farther and farther away from one another without changing their own size.

Hubble's discoveries force us to ponder some aspects of our own position in the scheme of things. Two specific questions arise:

Do we live at a special place? Why are the galaxies receding, in particular, from us here on Earth? Hubble's law seems to put our planet at a very remarkable place in the universe—at the center, a location apparently singular in its repulsiveness. Shades of Dante, who located Satan at the center of Earth, which Dante also considered the center of the universe.

Do we live at a special time? It is far from obvious that our present, uniform distribution of galaxies will not be spoiled by continuing expansion. Shouldn't the universe "thin out" at the center? For instance, if the velocities with which galaxies recede were all the same (instead of proportional to their distance), our descendants would find themselves at the center of a unique hole in the universe of galaxies. A distribution containing such a hole is evidently far from uniform.

Superficially, Hubble's law seems to suggest positive answers to these questions and therefore to assign a very special significance to the place Earth and the time Now.

On deeper reflection, however, we find that it does not. In fact, the precise form of expansion discovered by Hubble, with the velocity proportional to the distance—and no other form—turns out to look the same to all observers. An astronomer based on a distant galaxy would also see all the galaxies receding from him. We are therefore not in an especially repulsive spot; it's the same everywhere else.

Expansion à la Hubble also preserves uniformity. If at any one time the galaxies are uniformly distributed, they will remain uniformly distributed. So the present uniform distribution doesn't single out our time as special; our descendants will be able to admire a similarly uniform distribution of galaxies.

If these remarks seem paradoxical, consider this simple model. Imagine a ruler with little observers (sentient bugs) A, B, C, D, . . . , located respectively at the end, the 1-inch mark, the 2-inch mark, the 3-inch mark, ... Now, imagine that something causes the ruler to expand uniformly, doubling its size. Holding up a second, unexpanded ruler next to the first, we find that, according to the second ruler, A is still at the end, while B, C, D, . . . are at the 2, 4, 6, . . . marks. How does A, observing the other bugs, describe what has happened? He says that B started one inch away and receded one inch, that C started two inches away and receded two inches, that D started three inches away and receded three inches. . . . Each one has moved away by an amount proportional to his original distance. This recession of the bugs, in other words, has proceeded according to Hubble's law.

With this model in mind, let's reconsider our two questions. To answer the first question, let's examine how bug B describes what has happened. He says his nearest neighbor to the right, C, was originally one inch away but is now two inches away—he says C has receded one inch. Similarly, his second-nearest neighbor, D, was originally two inches away but has receded two inches. . . . This description of events should seem very familiar—in fact, it coincides exactly with A's description. Likewise, every one of the bugs sees the same pattern of recession à la Hubble. So we can answer our first question in the negative: that we observe all galaxies to be receding from us does not mean we are in a special position, because an observer in any other galaxy would report the same pattern of recession.

We also easily see, in this model, that recession following Hubble's law takes a uniform distribution of bugs into another, expanded, uniform distribution of bugs. And so we can answer our second question in the negative as well. The fact that we now observe a uniform distribution of galaxies does not imply that we live at a special time, because uniformity is preserved by Hubble's law.

Far from disturbing the uniformity of the universe, then, expansion following Hubble's law meshes with it beautifully. The great discoveries of modern observational cosmology allow us to stick with the most "radically conservative" assumption about our position in the cosmos—that we do not find ourselves at an especially peculiar place in the universe or at a peculiar time in its history.

6

Inferno and Afterglow

[T]he red-shift of spectral lines observed in distant galaxies must be interpreted as the consequence of a rapid expansion (or, rather, dispersal) of the far-flung system of galaxies populating the limitless space of the Universe. It follows that, once upon a time, the matter forming our Universe was strongly compressed possessing a, presumably uniform, high density and temperature. . . .
    —G. Gamow, *Vistas in Astronomy* (1956)

The expansion of the universe brings to our attention the question of its history. Having found that the universe is changing, we can't help wondering how it will look in the future and from what state it emerged.

In most fields of science, theories get judged by their ability to predict the results of new experiments and observations. (Is lightning electric? Let's fly a kite into a thundercloud and see if it brings down sparks.) Often, control of nature is the ultimate goal. (For instance, we learn to tame lightning with the help of lightning rods.)

Cosmology, the study of the universe as a whole, has a different flavor. Our prospects for doing controlled experiments in cosmology are extremely limited. We cannot set up different universes and see how they run. The experiment has been done; our universe is what it is. Scientific cosmology is an exercise in "postdiction"—in guessing the past.

It might seem that such an exercise cannot be scientific at all. How are we to test speculations about the distant past, since we cannot revisit it? One way is to search for relics. Just as paleontologists studying the history of life collect fossils, cosmologists look for artifacts that can be explained only as survivors from an earlier, very different time.

Remarkably, through the search for relics, postdiction often begets prediction. Paleontologists may find a gap in the fossil record of development of a species; they then predict that intermediate forms or "missing links" must exist, and set about searching for them. Detectives trying to reconstruct a crime are engaged in the same sort of enterprise. They search for a "smoking gun," a relic that will confirm their hunches about what happened. In each case, guesses about the past are metamorphosed into positive programs for research.

The search for relics from the early history of the universe is the core of scientific cosmology. But how do we get started? How do we know what to look for; how do we recognize such a relic when we find it? The approach that has proved fruitful is the radically conservative approach. In this spirit, we make the radically conservative assumption that the early universe was as simple as it possibly could have been, and draw the consequences of this assumption with all the rigor and completeness we can muster.

Beginning with these simplest-possible ideas about the initial condition of the universe, let us follow known laws of physics to find out what they imply about how it evolves. Let us see if something like the observed universe could have evolved from such an initial condition. If we are lucky, we will also find that the early universe should have left clear-cut relics. In scientific cosmology so far, the simplest-possible assumptions about the state of the early universe have proved to be tenable—and, luckily, to predict that some surprising relics should have persisted.

Since we are witnessing, in the recession of galaxies, something that looks like an ongoing explosion, it is reasonable to expect that there actually was an explosive, singular event at the start. (It sounds funny to be having expectations about the past, but that is what postdiction is all about.)

A more precise version of this intuition can be justified on the basis of the *time-reversal invariance* of physical laws. For our present purposes, the most useful way to understand the principle of time-reversal invariance is to compare our present, real world with an imaginary fantasy world—call it T-world. Right now, at this precise instant, T-world differs from the real world in just one respect. The two contain identical objects, identically arranged. The only difference is that every object in T-world has had its velocity reversed. The elevator going up the Empire State Building in our world is going down in T-world; cars moving north in our world proceed south (and backward) in T-world; in T-world, the hands of Big Ben rotate counterclockwise.

The principle of time-reversal invariance is the statement that the future development of the fantasy T-world, following the laws of physics, would reproduce the actual past of the real world. In other words, time-reversal invariance means that if we watch our fantasy world unfold for an hour (or a century), we will see it develop into precisely the state of our own, real world one hour (or one century) ago. The power of the principle of time-reversal invariance is that it allows us to <u>postdict</u> the past by means of the same techniques we have developed to <u>predict</u> the future. We just apply these techniques, in our imagination, to a suitable fantasy world.

To reconstruct the past of our universe, then, we should ask ourselves what would happen if the velocities of galaxies were reversed, so that the universe was contracting instead of expanding. The future of this fantasy universe (T-universe) is the past of our own.

It is helpful to think of the bugs on a ruler again. If the ruler shrinks, the bugs come closer together. As the distance between bugs vanishes, there is an infinite density of bugs everywhere. This model suggests that there was a big crunch in the past of our universe. Our universe expanded out of an earlier condition of very high density.

As gases expand, they tend to cool. You can demonstrate this yourself. Blow on your hand—first with your mouth wide open, then with pursed lips. The initial temperature inside your lungs is, of course, the same in both cases, but the breath forced through your lips at high pressure expands as it escapes, cooling in the process. Since the universe has been expanding and cooling for a long time, we can anticipate that it must have been much hotter in the past.

Another, more rigorous argument leads to the same conclusion. In T-universe, where velocities are reversed, the galaxies are coming closer together. As galaxies collide and merge, the energy associated with this overall motion of whole galaxies is distributed among many smaller bodies—stars, in the first instance—and partly randomized. In other words, part of this energy is turned into heat. The future of T-world is very, very hot, because part of the energy gained by objects as they "fall into" one another is transformed into heat. T-world's future duplicates our past, so we infer that our own universe must once have been extremely hot.

In our fantasy universe, the gravitational attractions between galaxies get stronger and stronger as they get closer. Because the force of gravity is greater at shorter distances, the galaxies are being accelerated—their velocities, and the energy in their motion, are increasing with time—at an ever-increasing rate. (In the standard language of physics, gravitational potential energy is turning into kinetic energy.) This behavior in the distant future of T-world—the more rapid contraction—implies the mirror image behavior (more rapid expansion) in the distant past of the real universe. The present rate of universal expansion has been greatly slowed down by gravity and is only the remnant of a more rapid, and at first explosive, expansion.

So the historical implication of the expansion of the universe is that the universe has evolved from a much denser, hotter, more rapidly expanding phase early on. The picture is reminiscent of an explosion, in which a small device is ignited and forms a fireball, which then expands and cools. For this reason, the picture of the universe we have started to draw is called big bang cosmology. This irreverent name for the start of the history of the universe is, in important ways, an appropriate description of its physical nature.

The large-scale distribution of galactic matter appears, we have said before, to be quite uniform. The radically conservative historical assumption, then, is that the distribution of matter has always been roughly uniform. The simplest-possible assumption is that the distribution in earlier times was more precisely uniform than it is today. Is this radically simple assumption tenable? Can the observed clumping of matter into planets, stars, and galaxies have emerged dynamically from a much more uniform distribution early on?

This idea has a chance to work because the force of gravity tends, in time, to exaggerate whatever nonuniformities there are. Gravitational attraction is proportional to mass, so its concentrating power works most effectively wherever mass is already highly concentrated. If one region of space happens to contain a higher density of matter than its neighboring regions, the destabilizing effect of gravity will cause it to become denser still. In time—for example, over the lifetime of the universe—this self-reinforcing property can greatly magnify any small initial density contrast. It therefore becomes possible to imagine that even very small deviations from perfect uniformity in the early universe are sufficient to serve as seeds from which such structures as stars and galaxies can grow, separated by vast empty spaces.

Combining clear-cut arguments with a radically simple guess has yielded a definite and striking model of the early universe. It was, according to this model, a hotter, denser place, and monotonously the same everywhere. In short, it was a structureless inferno.

If this model is accurate, then we should expect to find a remarkable relic—a much toned-down remnant of the big bang, which scientists call microwave background radiation.

To understand this radiation, it is easier to begin by thinking about the radiation from a very hot gas like that inside a neon light.

The same neon that goes into neon lights is, at room temperature (roughly 25 degrees Celsius), utterly transparent—more transparent, even, than ordinary air. Furthermore, the neon in the tube is much less dense than ordinary air. If the glass gets punctured, the tube implodes. Yet it is just this much-attenuated version of a transparent gas that gives Las Vegas its garish glow. How is this possible?

The answer is that the character of neon changes abruptly when it is heated up by an electric current passing through it. Indeed, the character of matter in general changes abruptly when it gets heated beyond about 3,000 degrees or so. Below this temperature, matter is electrically neutral, made up of atoms whose positively charged nuclei are exactly balanced by their negative electrons. At the high temperatures in a neon light, the electrically charged pieces of atoms become unstuck. Frequent and violent collisions break down neutral atoms into electrons and unbalanced nuclei. Matter in this state is called plasma, and it radiates much of its collision energy in the form of light.

In bound atoms, the canceling electrical charges of the protons and electrons are close together, and tend to move together, so that their interactions with light's electromagnetic field are effectively canceled. In other words, a gas of neutral atoms (like air) is virtually transparent. The free nuclei and electrons of plasma, by contrast, couple to light's electromagnetic fields and absorb it very efficiently.

Now, imagine looking at a big hot ball of neon, like an artificial sun perhaps, hottest at the center and cooling off toward the outside. You would see light only from the borderline layer of neon between opaque plasma and transparent neutral atoms—neon at a temperature of close to 3,000 degrees.

This big ball of neon is the image, in spatial terms, of a complex process in time. According to big bang cosmology, the universe also has a changing temperature—one changing in time rather than in space. Just as the temperature in and around a flame decreases as you get farther from its center, so the temperature of the universe decreases with time, as it expands and cools. The expansion of the universe has cooled it enormously since the early stages of the big bang.

At very early times, very hot material was everywhere—every bit of matter in the universe was in the form of plasma, emitting enormous amounts of light but also immediately absorbing it again. That light is gone, and we don't get to see it. Then the material cooled down; when the temperature fell below roughly 3,000 degrees, it became transparent. The light from that border time is still visible to us.

The universe as a whole is now a very transparent place, letting light through to us from enormous distances across time as well as space. By measuring distances in light-years, we remind ourselves how closely time and space are bound together. The farther out into space we look, the farther we can see into the past. We can see light from the sun as it was eight minutes ago, light from the neighboring Andromeda galaxy emitted two million years ago, and light from the plasma borderline of the big bang emitted ten billion years ago. There is a tremendous amount of big bang light, because at very high temperatures like those in a neon lamp, lots of light gets emitted— especially if it's the entire universe everywhere doing the emitting.

Suddenly, our relic seems like too much of a good thing: Why isn't the entire sky glowing like Las Vegas, lit up by residual light from the big bang?

Light from the big bang has been traveling through space since close to the beginning of the universe; we are now seeing light from ten billion light-years away. The velocity of its source away from us is very close to the speed of light, and the Doppler shift decrease of its frequency is correspondingly enormous, a factor of a thousand. This light has been shifted so far toward the red that it has left the visible spectrum altogether. It has metamorphosed into a form of radiation called microwaves, which the eyes we are born with do not sense (microwave radiation is, of course, the form of radiation used in microwave ovens). It has become a pale fire, a lingering afterglow from the receding big bang.

In volume 142 (1965) of the *Astrophysical Journal,* a very brief paper with the modest title "Excess Antenna Temperature at 4080 Megahertz," by Arno Penzias and Ralph Wilson, appeared. It begins as follows:

> Measurements of the effective zenith noise temperature of the 20-foot horn-reflector antenna at the Crawford Hill Laboratory, Holmdel, New Jersey, at 4080 Mc/sec. have yielded a value about 3.5 K. higher than expected. This excess temperature is, within the limits of our observations, isotropic, unpolarized, and free from seasonal variations. . . .

In this deadpan prose, one of the great cosmological discoveries of all time was announced to the world. Without seeking it, and without at first realizing what they had found, Penzias and Wilson had discovered the relic radiation from the big bang. Although its existence had been predicted as early as the 1940s, these "wild" predictions had been generally ignored and were unknown to Penzias and Wilson.

A microwave antenna looks basically like an old-fashioned ear trumpet, blown up to gigantic size. Pointing one in different directions makes it possible to compare how much microwave radiation flows into it—in scientific terms, the flux of radiation—from different parts of the sky. This flux, to an amazing degree of accuracy, turns out to be the same in every direction. In other words, if our eyes responded to microwave radiation instead of to visible light, then when we looked at the night sky we would see not stars but an utterly featureless, omnipresent haze. We say the flux is *isotropic.*

A featureless haze is not a rich mine of information. In an effort to extract more, ever since the discovery of the microwave background radiation major experimental efforts around the world have been mounted to tease out some anisotropy, some nonuniformity in the flux from different parts of the sky.

There is one way in which the background departs from perfect isotropy. We appear to be moving through the microwave background, at a velocity of roughly 520 kilometers per second, in the direction of the constellation Hydra. This motion leads to a Doppler shift of the microwave frequencies observed. Microwaves arriving from the direction of Hydra are shifted toward higher frequencies; those from the opposite direction, toward lower frequencies. (Our velocity, by the way, is typical of the so-called <u>peculiar</u> velocities of galaxies with respect to one another, random velocities that exist in addition to the systematic recession of distant galaxies found by Hubble. Here again, we find ourselves in an undistinguished position; observers in other galaxies would find quite similar anisotropies.)

Aside from this effect, the search for anisotropy in the microwave background has succeeded only in revealing with greater and greater precision just how uniform that background is. It is now known that the microwave flux from differing directions in the sky is the same, to an accuracy of one part in ten thousand.

The isotropy of the microwave background, though frustrating to observers, provides dramatic confirmation of our radically conservative guess that the early universe was extremely uniform. If matter had been clumped when the microwaves were emitted, this "light," like the ordinary light from stars and galaxies, would reveal the patchy pattern of its sources.

In confirming the radically conservative hypothesis—that the early universe was even more uniform than our own—the isotropy of the microwave background presents us with a challenge and an opportunity. As we've indicated, even tiny deviations from uniform density will grow automatically, under the influence of gravity. The large-scale structure of the present universe is very likely encoded in tiny fluctuations at the time when the microwave background was emitted, fluctuations whose signature we expect to find written in tiny departures from the perfect isotropy of this background.

If we can figure out what the small seed fluctuations were like, it will most likely be possible to understand the subsequent development of structure in the universe by pure calculation, by calculating how gravity causes the seeds to grow. With this motivation, the challenging search for tiny anisotropies in the microwave background continues to fascinate cosmologists. Important new data may come from the space satellite COBE (Cosmic Background Explorer), to be launched in the late 1980s.

Three Ages

We know of three very different ways to measure the age of the universe. Each leads to the same conclusion, that the universe is between ten and twenty billion years old. This age is of considerable interest in itself, since it sets the scale of time over which astronomical and biological evolution (both terrestrial and, possibly, extraterrestrial) has worked. Equally important is the very fact that all three determinations of the age agree. Their agreement gives us confidence that our extrapolations of physical laws into the remote past, and our radically conservative reconstruction of what the universe was like then, are not wildly wrong.

Before we proceed further, it may be worthwhile to consider a question that disturbs many people: If the universe has a finite age, what happened before? The honest answer may seem craven but actually contains the heart of the matter: we don't know. What we mean when we say that the universe is somewhere between ten and twenty billion years old is that back then the universe was a very different and violent place and that <u>few if any relics of what happened before could have survived.</u> We don't know about what happened before, because we can't: as far as we know, all traces of it have been obliterated. Universal history, in the sense of events that influenced the universe we know, began ten to twenty billion years ago.

It may be that someday people will be able to recognize some relic of events before the big bang. Then it would be necessary to contemplate earlier times and a longer "age of the universe." Even if this happened, the occurrence of the big bang, ten to twenty billion years ago, would remain an important turning point in the history of the universe. It would probably still be convenient to call it time zero and to divide cosmic time into B.B.B. and A.B.B., analogous to B.C. and A.D. In this language, when we measure the age of the universe, we are attempting to determine what time it is A.B.B.

The simplest way to measure the age of the universe is to project its present expansion backward in time. We can reverse the expansion, in our imagination, and ask how long ago it started. In other words, how long has it been since the big bang? From the work of Hubble and his successors, we learn the present distances and velocities of the distant galaxies. If the velocity of each galaxy were unchanging, it would be a simple matter of division to calculate how long it took them to get as far away as they are. We would find the same time for every galaxy, because the velocity is proportional to the distance (Hubble's law). We would be entitled to call this the age of the universe; more precisely, it is the time since the universe emerged from a much hotter, denser state.

In fact, the velocity is not quite constant. The recession of galaxies is continually slowed down by gravity, because the galaxies attract one another. If we knew the density of the universe, we could calculate just how big this effect of gravity is. Unfortunately, the density of the universe is still a controversial issue among astronomers. Also, there is some disagreement about just how far away distant galaxies are. Because of these uncertainties, we cannot derive an exact figure for the age of the universe from its rate of expansion. We can pin it down only roughly, as being somewhere between ten and twenty billion years.

A second way to measure of the age of the universe is to measure the age of its oldest atomic nuclei. The principle of this method, called radioactive dating, is very simple. Suppose nucleus A spontaneously changes into nucleus B, with a half-life of one billion years. This means that a sample initially containing only A will contain half A and half B after one billion years. After two billion years, half the remaining A will have decayed, leaving one-quarter A and three-quarters B. The content of the sample will tell its age. More exactly, it will tell us when the sample consisted of pure A.

(While the basic principle of radioactive dating is simplicity itself, its application to real materials can be a challenging art. Everything hinges, in our example, on starting with a sample of pure A. If the amount of B present initially is not known, the method becomes useless. Fortunately, there are many tricks, and variations on the basic idea, that help us get around this problem.)

Such nuclear "clocks" are extremely rugged. You can heat uranium in an oven at many thousands of degrees, or subject it to the largest available pressures, without affecting at all the rates of spontaneous nuclear decay. Therefore these clocks are well suited to dating (for example) meteorites, even though the meteorites have been exposed in interplanetary space to the depths of cold and have impacted on Earth after partly burning up in the process of crashing through its atmosphere.

When the techniques of radioactive dating are applied to meteorites, a remarkable result is obtained. All the meteorites condensed between four and five billion years ago. This tells us when the gas cloud that gave birth to our solar system was becoming dense enough that large numbers of molecules could find one another and stick together to form solid bodies of respectable size. It is not unreasonable to call this time the origin of the solar system. If we do so, we find that the age of the solar system is between four and five billion years. This age is a respectable fraction of the age of the universe, as inferred from its expansion, but definitely smaller.

Given a theory of element formation, the same ideas can be used to determine the age of our galaxy. Essentially all the elements except hydrogen and helium are thought to be produced in supernova explosions. (We'll be discussing the origin of the elements in theme 4.) Two forms of uranium—^{238}U, containing 92 protons and 146 neutrons, and ^{235}U, containing 92 protons and 143 neutrons— are assumed to be produced about equally in supernova explosions. But the ^{238}U is much more common today. The obvious explanation, given that the ^{235}U decays with a half-life of only (!) seven hundred million years, is that most of the ^{235}U has been lost to decay since its formation.

If the supernova explosions producing both kinds of uranium had all occurred at once, it would have taken seven billion years of decay to reach the present relative abundance. However, we know that supernova explosions continue to occur, so this "age" should be taken with a big grain of salt.

To define a more meaningful age, we need a more realistic guess about how many supernova explosions occur at various times. One popular idea is that there was a big burst of star formation, and supernova explosions, when our galaxy first condensed, followed by a more or less constant rate since then. This model yields an age for our galaxy of somewhere between eight and fifteen billion years. Although the model is no doubt oversimplified, it cannot be very far off. It has been successfully tested by checking its predictions for the abundance of other unstable isotopes. So we can conclude, with considerable confidence, that our galaxy is between eight and fifteen billion years old.

What does the age of our galaxy tell us about the age of the universe? As the universe continues to expand, the conditions for galaxy formation become less favorable. The density is decreasing, and it gets harder for gravity to condense lumps out of the ever-thinner gruel. Most models of galaxy formation predict, therefore, that our galaxy is only slightly younger than the universe itself. It is reassuring that the age of the galaxy determined by radioactive dating and the age of the universe determined from its expansion are closely comparable.

A third measure of the age of the universe involves the age of its oldest stars. We can calculate, for stars of different masses, how long they burn hydrogen before exhausting this fuel. It turns out that the stars with the least mass survive longest. (Such stars have less fuel, but they burn it much more slowly.) Astronomers look for the most massive stars that are still burning hydrogen. Stars slightly more massive than these, we infer, have just recently burned out. Since the calculations tell us their expected life span, the knowledge that they just died allows us to infer their date of birth. Carrying out this program, we find that the oldest stars are between ten and fifteen billion years old.

So there are three very different ways of determining the age of the universe, and all agree within their uncertainties. This accord gives us considerable confidence in the result. No less important, it gives us confidence that the physical laws underlying the calculations have not changed much for a very long time.

Third Theme

Transformations

From Thales on down, the classical approach to change in the physical world has been to explain it as being merely the rearrangement of more fundamental entities, themselves unchangeable. Again and again, though, supposedly changeless elements of reality have turned out to be changeable after all. From the supposedly perfect and inalterable bodies "beyond the moon" to atoms, nuclei, and even protons and neutrons, changes and transformations have been found in one "immutable" object after another. It now seems that only much more abstract and intangible things, such as energy and electric charge, truly persist unchanged in time.

The omnipresence of change, and the primacy of transforming principles it suggests, may be psychologically discomfiting. It discredits our natural hope that tangible eternal realities underlie the appearance of worldly mutability.

On the other hand, something wonderful happens when we shift our intellectual allegiance from <u>things</u> to <u>laws of transformation</u>: the content of the world, no longer something we must simply accept as given, becomes a legitimate subject for scientific investigation.

TREIMAN'S THEOREM

Sam Treiman of Princeton University is famous not only for his original work in theory of elementary particles but also for his general wisdom. On many occasions, he has quoted something he called Treiman's theorem. Treiman's theorem is a bit of profound nonsense. It is deceptively simple to state, as follows:

Theorem: Impossible things usually don't happen.

Sam invariably quoted his theorem when some startling new experimental development was announced. Since it is all too easy to make an error somewhere along the line in constructing, running, or analyzing experiments of the subtlety and complexity now standard in high-energy physics, Treiman's theorem was often successfully applied. Claims of very surprising, "impossible" results usually did not survive. Careful scrutiny or repetition of the experiment cast doubt on the original, startling results.

Properly interpreted, however, Treiman's theorem is much more than a simple warning against credulity. Taken literally, of course, it is a silly understatement—impossible things, by definition, are a step beyond being unusual. But by juxtaposing these two ideas, by hinting that, well, just maybe, impossible things do after all occasionally happen, the theorem strikes a deeper chord. It reminds us of the humble origins of most claims that something is impossible. Apart from logical contradictions—it is impossible to have a four-sided triangle—something is said to be impossible simply because it has never been seen or, in other words, precisely <u>because</u> it usually doesn't happen.

A notable example, very relevant here, is the doctrine of chemical elements. Order was introduced into chemistry during the eighteenth century, when it was demonstrated by many examples that chemical substances could all be synthesized from a small number of basic substances, or elements. For example, water can be broken down into hydrogen and oxygen, and rocks (which are much more heterogeneous than water) can eventually be broken down into silicon, aluminum, oxygen, and traces of other elements, including—in favorable ores—gold and silver. The elements, on the other hand, resisted all attempts to break them down further or to change one into another.

Thus chemical molecules are understood to be groupings, in definite proportions, of atoms of the elements. In chemical reactions, the groupings can change, but the elementary atoms do not. Perhaps the most important aspect of the doctrine of elements is that it defines the chemically impossible. For example, the goals of the alchemists—to transmute mercury into silver or to create gold out of substances that do not already contain it—are "impossible," because mercury, silver, and gold are all chemical elements. The great achievement of the doctrine of elements is to draw a line separating possible reactions (those that can be achieved by rearranging elementary atoms) and impossible ones (everything else).

With the discovery of radioactivity at the beginning of our own century, it suddenly became apparent that the "impossible" was happening all the time. Uranium, thorium, radium, and a few other substances were found that fit all the requirements for chemical elements. They could not be broken down by any of the standard methods—heating them, passing electric currents through them, exposing them to acids, and so forth. But occasionally, in a process that could be neither halted nor speeded up by any of these methods, atoms of these elements spontaneously changed into other kinds of atoms. Uranium, for instance, can undergo a complex series of changes whose end products are several atoms of helium and one of lead.

So what is left of the doctrine of elements? Is alchemy reinstated? Not at all. The point is that the doctrine fails only under rare or special conditions. Impossible things usually don't happen. We can isolate the conditions in which they do, and retain a more restricted but still useful concept of the "impossible."

This example is far from isolated. For instance, it is a useful fiction to say that different atomic nuclei are made of protons and neutrons and that in nuclear reactions these "elements" are rearranged but not themselves transformed. In fact, however, there are processes by which neutrons and protons can change into one another. In the simplest case, an isolated neutron lives only about ten minutes before decaying into a proton, an electron, and an antineutrino. But there is a real sense in which these processes of transformation (the so-called weak interactions) are rare and unusual compared with simple rearrangements of protons and neutrons (the strong interactions). Reflecting this disparity, the typical nuclear rearrangement reaction occurs some 10^{22} times faster than the decay of a neutron. Evidently, the weak interactions are much less potent than the strong. We can usually ignore them, in which case it becomes "impossible" to change the number of protons or neutrons in a nuclear reaction, just as it is "impossible" to change the number of atoms of any element in a chemical reaction.

In each of these cases, and in several others we shall encounter, Treiman's theorem at its deepest level defines a program for investigation. We seek, in organizing a reality marked by complex transformations, ever to refine our notion of the impossible. In chemistry, the alteration of elementary atoms is impossible; in strong nuclear interactions, the alteration of protons and neutrons is impossible. In neither case is the constraint on reality absolute—it's just that in some well-defined circumstances the impossible usually doesn't happen. Naturally, the unusual processes that do occasionally make the "impossible" happen then become especially interesting. These processes of transformation slowly modulate the otherwise fixed elements out of which we construct our description of the world.

7

New Star

I cannot without great wonder, nay more, disbe-
lief, hear it being attributed to natural bodies as
a great honour and perfection that they are im-
passible, immutable, inalterable, etc.: as, con-
versely, I hear it esteemed a great imperfection to
be alterable, generable, and mutable. . . . These
men who so extol incorruptibility, inalterability,
and so on, speak thus, I believe, out of the great
desire they have to live long and for fear of death.
. . . These people deserve to meet with a Medusa's
head that would transform them into statues of
diamond and jade, that so they might become
more perfect than they are.
  —Galileo, *Dialogue on the Great World*
  Systems (1632)

The longing for permanence is among the deepest instincts that lead
men to religion, and also to philosophy and to science. This instinct
is no doubt linked to some of our most basic and most deeply felt
biological needs for stability and refuge from danger—and to our
fear of death.

Medieval Scholastic philosophy and theology, following Aristotle,
claimed permanence and perfection for certain parts of the physical
world. Earth is manifestly a region of change and imperfection. It
was held, though, that the moon and the sun, planets, and stars
beyond had to be of an entirely different nature, unchanging and
perfect. The moon, for example, could only be a sphere, since this is
the only "perfect" shape.

In 1572, a dazzlingly bright new star appeared in the sky, near the constellation Cassiopeia. (Today we would call it a supernova.) This supernova came at a very fortunate time, amid the intellectual ferment of the Renaissance. A few people were interested enough in the natural world, and confident enough in their own power to observe what was actually happening—whether or not it was what the authorities said was supposed to be happening—to take it very seriously.

Among these was Tycho Brahe. Tycho was a wild character, an alienated Danish nobleman who had lost part of his nose in a duel with a rival and wore a replacement made of silver and gold. The dispute leading to the duel had started over a disagreement regarding who was the better mathematician. Tycho as a fourteen-year-old student was deeply impressed by a partial eclipse of the sun—not so much by the event itself as by the fact that its occurrence had been predicted. Three years later, he had his second formative experience, which built, in an ironic way, on the first: he witnessed a conjunction of Saturn and Jupiter that took place a month later than the standard planetary tables predicted. He considered this a shocking state of affairs and decided to dedicate his life to improving the quantity and quality of astronomical observations, by watching the sky full-time with improved instruments. Using family funds and the royal gift of a small island near Copenhagen, he built the world's greatest, and last, observatory for naked-eye astronomy.

When the new star appeared, Tycho was ready to answer a key question about it: Was it farther away than the moon? If it was, the doctrine that no change occurred beyond the moon would fall. Tycho was well equipped to see that the new star, unlike the moon, remained absolutely fixed in its position relative to the background stars. Except for its obvious mutability (after its dramatic entrance, it proceeded to fade back to invisibility over several months), the new star behaved in every way like a regular star.

Tycho's painstaking observations showed that the new star had to be farther away than the moon. With this, the medieval and Aristotelian notion of a domain of the physical universe immune to change began to crumble.

Not long afterward, in 1609, Galileo used the newly invented telescope to observe features on the moon that showed that it "is not perfectly smooth, free from inequalities and spherical . . . on the contrary, it is . . . just like the surface of the earth itself, which is varied everywhere by lofty mountains and deep valleys." Here was an observation anyone with eyes could repeat, a visible refutation of the medieval ideas. The reaction of some "establishment" scholars was to refuse to look. On the death of one of these, Galileo unleashed a fine bit of sarcasm: "Libri did not choose to see my celestial trifles while he was on earth; perhaps he will do so now he has gone to Heaven." Humor could not fully conceal the stakes involved, however. Compare this sally with the epigraph for this chapter. And try to imagine, on the other side, the traumatic emotional impact of the sudden collapse of a complete worldview, which gave comfort and security to those who accepted it and which was endorsed by centuries of tradition and by the authority of the church.

The idea that astronomical objects must be perfect or permanent never recovered from the blow that Galileo dealt it. Indeed, we have seen that two major themes of modern physics are that there is no fundamental distinction between earth-stuff and star-stuff and that the universe is evolving. Modern physics has recently come to recognize that the fundamental interactions governing the behavior of elementary particles include processes of transformation; we shall elaborate on this development in the next chapter.

As the idea of permanence of objects has faded, the idea of permanence of physical laws has become better established and more powerful. We can check in a remarkably direct way that physical laws are not changing. The light reaching us from distant galaxies was emitted hundred of millions of years ago, yet it shows the same patterns of spectral lines that we elicit from matter today. The forces that determine the shape of atoms and molecules, and thereby the frequencies of light they resonate with, have not changed significantly in all this time.

An increase in the predictive power of physical laws is the flip side of the demise of permanence of objects. If no objects are permanent, then the origin of all the objects we actually find in the world must be understandable as a consequence of the working of physical laws in history. This means that the form of physical laws not only tells us how the universe will evolve in the future, but also plays a major role in determining what we find in it today. The traditional division between initial conditions ("what there is") and laws of motion ("how things work") is crumbling.

8

The Weak Interaction

*. . . God hath chosen the weak things of the
world to confound the things which are mighty;
And base things of the world, and things which
are despised hath God chosen, yea, and things
which are not, to bring to nought things that are.*
—Saint Paul, 1 Corinthians 1:27–28 (53 A.D.)

If the main task of science is to make predictions or, more generally,
to understand how the world unfolds in time, it must identify precisely what it is about the present that carries over into the future.
A simple guess, which for many centuries proved adequate and fruitful, is that certain <u>substances</u> persist unchanged through time.
Democritus, often considered the father of atomic theory, put it this
way:

> By convention sweet is sweet, by convention bitter is bitter, by convention
> hot is hot, by convention cold is cold, by convention color is color. But in
> reality there are atoms and the void. That is, the objects of sense are
> supposed to be real and it is customary to regard them as such, but in
> truth they are not. Only the atoms and the void are real.

Atomic theory, in its classic form, ascribes reality—permanence—
to certain primordial units of matter. According to this theory, matter is composed of small indivisible, unchanging units—atoms, from
a Greek word meaning "indivisible." All physical change is to be
understood as merely the rearrangement of atoms.

A more modern version of the program of atomic theory was
expressed by Hermann von Helmholtz (1821–1894) in the paper in
which he first formulated the general principle of conservation of
energy:

Now in Science we have already found portions of matter with changeless forces (indestructible qualities) and called them (chemical) elements. If, then, we imagine the world composed of elements with inalterable qualities, the only changes that can remain possible in such a world are spatial changes, i.e. movements. . . . So that at last the task of Physics resolves itself into this, to refer phenomena to inalterable attractive and repulsive forces whose intensity varies with distance. The solubility of this problem is the condition of the complete comprehensibility of nature.

In pursuing this program, we try to understand all the myriad changes we observe as rearrangements. For instance, the transformation of water into ice is understood as a packing of its molecules into orderly crystals, and chemical reactions are understood as regroupings of atoms. All rearrangements, of course, are caused by motion, and motion in turn is governed by forces. Thus the classical program for physics was to reduce the description of the world to a small number of forces acting upon a limited inventory of unchangeable atoms.

Recently, physics has been forced to abandon the idea of permanent substances, of material units that persist unchanged in time. To appreciate the depth of the change in our worldview that abandoning the ideal of permanence implies, it is helpful to review its downfall.

The first blow came when atoms themselves, the smallest units of the chemical elements, were found to have parts after all. They are composed of electrons and nuclei. This discovery in itself requires only a trivial change in the basic program—namely, the substitution of electrons and nuclei for atoms as the basic indivisible units. The fundamental idea that there are unchanging, indivisible units still did not really come into question.

A second blow came when it was found that in radioactive decay the nuclei themselves can be transformed. Many people still held on (in the true radically conservative style) to the idea of permanence. They hoped that the recent history of atomic physics would be repeated for nuclei, that the appearance of divisibility and change was in reality merely rearrangement of smaller, permanent parts. When James Chadwick discovered the neutron, in 1932, it seemed that their faith had been rewarded. All the different nuclei could be understood as being composed of different numbers of protons and neutrons bound together. The mass of any nucleus is roughly the sum of the masses of the neutrons and protons it contains. Transformations among the nuclei could be understood as rearrangements of their parts, the protons and neutrons.

At this point, you may be becoming suspicious. Is the program of Helmholtz no more than an empty shell, capable of accommodating any observation? Is permanence a law of nature—or a law of thought—or simply a failure of imagination? When protons and neutrons are found not to be permanent, wouldn't the inevitable reaction of physicists be to postulate still smaller objects lurking underneath?

In fact, this was the reaction of most physicists to the discovery that neutrons can spontaneously decay. It soon became apparent, however, that something basically new was happening, something that would not fit into the old framework. A new theme entered fundamental physics: the theme of transformation.

As we have seen, the discovery that an isolated neutron will break down does not necessarily violate the ideal of permanence. To save this ideal in the traditional way, one would simply argue that the neutron is "made of" its decay products, that its decay is simply what happens when these permanent particles come unstuck. This traditional response proves inadequate, however, when we consider neutron decay in more detail. The products of the decay are a proton, an electron, and an antineutrino, so we write

$$\mathbf{n} \rightarrow \mathbf{p} + \mathbf{e} + \mathbb{V}$$

where $\mathbf{n}$ = neutron, $\mathbf{p}$ = proton, $\mathbf{e}$ = electron, and $\mathbb{V}$ = antineutrino. The proton and electron are familiar constituents of ordinary matter, represented straightforwardly by their initial letters, but the antineutrino (represented by a Greek letter *nu* in outline type) is quite something else and requires further explanation. Actually, it was invented by Wolfgang Pauli, in this very context. Pauli was making a bold and seemingly desperate attempt to save the conservation laws of physics. When neutrons decay, the observed decay products are just a proton and an electron (and nothing else). But the total energy of the observed decay products, the proton and the electron, does not match the total energy of the original neutron. Likewise, the total momentum of proton plus electron differs from the original momentum of the neutron. (A particle's momentum— roughly speaking, its mass times its velocity—has been used as a measure of its conserved quantity of motion since the time of Galileo and Newton.) If energy and momentum are to be conserved, however, the energy and momentum of the decay products must be precisely the same as those of the starting neutron. Pauli was very reluctant to give up either of these conservation principles, which had been successfully tested in many different contexts and were deeply woven into the fabric of theoretical physics. He suggested instead that the observed decay products are not the only decay products. He proposed that an additional particle is emitted, one that carries off the "missing" energy and momentum but is not easily observed directly. The particle in question came to be called the neutrino, Italian for "little neutral one." (The *anti* in *antineutrino* is another story. We shall discuss the whole subject of antimatter, of which this is one example, at some length in chapter 18.)

Pauli made his proposal very tentatively and never published it. It was included as almost a mocking jest in some letters he wrote to friends. Why was he so coy?

For one thing, Pauli was very proud of his reputation for correctness, and he could be extremely sarcastic about other people's errors. He later became known as the conscience of physics. He evidently did not feel confident enough about his neutrino proposal to put it forward in the normal scientific literature. Of course, Pauli knew that his friends could be relied on to get the word out anyway.

More important, perhaps, was a certain philosophical unease about what he was doing. In one of these letters, he said, "I have committed the ultimate sin, I have predicted the existence of a particle that can never be observed."

Most important of all was probably Pauli's realization that his new idea clashed with the old ideals of permanence. Having unconsciously adopted the traditional idea of permanent entities, he thought that if a neutron decays into proton, electron, and antineutrino, then it must have contained those particle to begin with. There are, however, as Pauli immediately perceived, severe difficulties with the idea that a neutron is "really" a proton, electron, and antineutrino bound together.

Here is the problem. Protons and electrons bind together into hydrogen atoms. A hydrogen atom compared with a neutron is enormous, its radius being almost a million times larger. If there were powerful short-range forces that could bind together protons and electrons into the small volume of a neutron, these forces would make hydrogen atoms collapse to the size of a neutron. Perhaps the antineutrino plays a critical role in the neutron? That notion won't work, because the very essence of the antineutrino idea is that this particle escapes detection. For this to be true, the forces it exerts on other particles must be very feeble. In short, there just aren't any forces available to hold the proton, electron, and antineutrino together. A neutron cannot be formed from these constituents. The decay of a neutron cannot be understood as the rearrangement of these parts.

Enrico Fermi adopted Pauli's proposal that a particle was escaping detection in the process of neutron decay, but he renounced the idea that this particle had been hiding in the neutron beforehand. In other words, Fermi escaped the bewitching spell of the idea of permanence. Renouncing the attempt to supply a detailed model of the process, he simply postulated, on the basis of the undeniable fact of neutron decay, that <u>processes of transformation, as well as forces, must be taken as fundamental laws of physics</u>. In other words, he proposed that the laws of physics must include not only rearrangements of some permanent underlying stuff but also rules for transformations that could not be interpreted this way. In the case of our neutron decay, Fermi's idea means that we can measure some finite probability (per unit time) that a neutron will disappear and be replaced by a proton, electron, and antineutrino at the same point—period. There is, according to Fermi, no new <u>force</u> among these particles, just this new <u>transforming interaction</u>.

So far, we have described Fermi's achievement rather negatively, as a renunciation and as a resisting of bewitchment. It is much more than a renunciation, however, and leads to important, positive, testable results. Most immediately, it gives a prediction for the energies and angles at which the observable protons and electrons in neutron decay are emitted. If we had to worry about complicated unknown forces, of course, it would be impossible to predict the trajectories (paths of flight) of these decay products. Once Fermi's radical idea—that there simply <u>are</u> no forces—is accepted, the trajectories can be figured out mathematically. The calculated trajectories agree closely with those observed.

Fermi's idea of transformations as fundamental laws of physics has surprising consequences of another sort. Neutron decay is not an isolated case. Once this transformation is allowed, a whole series of other transformations is inevitably implied. For one thing, you can reverse the arrow—the inverse process

$$p + e + ʊ → n$$

must also be possible. (As a practical matter, this process is difficult to observe, because it is difficult to bring the three reactants together.) Also, you can move a particle from one side to another, if at the same time you replace it by its antiparticle. These basic operations can be combined. For example, from the original process

$$n → p + e + ʊ$$

you get the possibility of

$$n + ν → p + e$$

where $ν$ = neutrino, by moving the antineutrino to the other side and replacing it with a neutrino. The implication is that a neutrino is capable of transforming a neutron into a proton and an electron.

An antineutrino can transform a proton into a neutron and an antielectron (also called a positron):

$$p + ʊ → n + ⊖$$

where ⊖ = antielectron. The existence of this reaction is further proof that a neutron is not "really" hiding a proton inside itself. We can "get to" this reaction from our original neutron decay, this time by changing the direction of the arrow <u>and</u> swapping the electron on one side of the arrow for its antiparticle on the other.

With all these new possibilities, the subject of neutron decay has itself been transformed—into something immensely richer. It has developed into a bigger subject of transforming interactions. This whole subject, to which neutron decay gave birth, is called the theory of *weak interactions.*

Weak interactions like those described above, whereby neutrinos and antineutrinos change neutrons into protons and vice versa, can set off alchemical changes in nuclei. Such processes were finally observed more than twenty years after Fermi proposed his theory.

The main difficulty in observing them is simply that the "little neutral ones" interact so feebly. Ordinary matter is essentially transparent to neutrinos. The antineutrinos emitted in neutron decay, for example, would on the average pass through ten million Earths laid end to end before being absorbed. A neutrino moving through matter shouldn't be compared to a speeding bullet plowing through a wall, leaving a trail of destruction in its wake. Rather, a neutrino is a ghost-like particle that passes through matter as if the matter weren't there and that leaves no trace. Neutrinos penetrate huge quantities of matter not because they have the power to shove things aside but rather because their interaction with matter is so feeble that matter presents practically no obstacle. The word *weak* in weak interactions derives from this extraordinary feebleness.

It is only in circumstances where very large numbers of neutrinos are produced—in a nuclear explosion, or near a nuclear reactor— that the tiny probability that any one neutrino will be detected is made up for by sheer weight of numbers. Out of a vast swarm of neutrinos, a very few will interact inside a large chunk of matter. If the right instruments are installed within the chunk, these reactions will be detected, leaving their signature in the changes they induce and the energy they deposit. By such means, Pauli's "sin" of predicting the existence of an undetectable particle has been redeemed.

It is fun to indulge in a little psychological speculation about how an intellectual giant like Pauli could come so close to developing a successful theory of neutron decay, and to discovering the whole theory of weak interactions, only to shy away from the final crucial step. He had both the basic idea of the neutrino and the technical equipment to exploit it, but instead he treated it like a joke. We are convinced from reading the literature as well as from personal experience, that there is a diversity among scientific personalities no less pronounced than the diversity among personalities as commonly understood. Pauli, in his scientific personality, was the conservative, the skeptic, the mocker, the purist. He was comfortable upholding the traditional conservation laws in the face of apparent violations; but the same intellectual conservatism made it psychologically impossible for him to abandon the traditional notion of the permanence of matter. (It is a fantastic, fascinating fact that Pauli, the scientific archconservative, became deeply interested in Jungian psychology and the mysticism of the Kabala. "Do I contradict myself? I contain multitudes.")

Fermi was a very different sort. He was an earthy, pragmatic man, one of the very few important modern theoretical physicists who was also capable of doing experimental work of the highest order. Fermi was not the type to hesitate in trying any idea that came to hand, or to be held back by the peculiar inhibitions and scruples that so troubled Pauli. On the other hand, Fermi was uncomfortable with a theory that could not be made to yield concrete results, that could not produce definite numbers to be compared with experimental measurements. In accordance with his scientific personality, he lost very little time in making the vague neutrino theory take definite shape and in putting it to the test.

Another psychological factor should be mentioned. The inventor of a genuinely new, promising idea often feels a pride in his discovery and a related affection for the idea, which makes him fear for its health and safety. He may prefer to leave the idea in a vague and flexible form, so that it runs no risk of contradiction by unfortunate experimental results. Naturally, his colleagues are not handicapped by such feelings, and it is they who are then able to take the idea further.

We have described how the new theme of transformations entered fundamental physics. Let us now discuss its wider meaning. For physics itself, the theme of transformation, as embodied in the theory of the weak interaction, provided both a unification and a paradigm.

The unification is that a long-standing dichotomy between matter and light was removed. When light is produced, a charged particle, such as an electron, is transformed into an electron and radiation—or, in other words, an electron and a particle of light (photon). Absorption is the same interaction, in reverse. From this point of view, impermanence did not begin with the weak interaction; it had long been accepted that light could come into being and pass away. The example of light was, however, almost too familiar. People took it for granted or regarded it as a thing set off by itself, not to be compared with (permanent, imperishable) matter. The weak interaction teaches us that "matter"—specifically, neutrons, protons, and electrons—can be as evanescent as light. In doing so, it hints that these elements of the physical world, matter and light, which to our senses seem so different as to be incommensurable, might after all have a single, unified description.

The new paradigm is that rules of transformation replace push-pull forces as the foundation of fundamental physics. We shall see in later chapters how this new paradigm has grown to dominance.

The theme of transformation in physics should be appreciated as part of a broader movement in intellectual history, away from a naïve way of thinking that comes to us all too naturally. At the beginning of this chapter, we quoted Helmholtz, who claimed that understanding the world in terms of permanent objects being rearranged in space is "the condition of the complete comprehensibility of nature." This style of explanation is indeed deeply ingrained in our nature. In the process of visual perception, we generally interpret a two-dimensional jumble of light impinging on our retinas as a sign of solid objects arranged in three-dimensional space. When dealing with other people or avoiding danger, we must learn to interpret and anticipate the motion of certain of these solid objects. It is very convenient—intellectually necessary, really—to assume that there is a high degree of permanence in the objects we are dealing with and to organize our concepts in terms of permanent objects moving through space. It should come as no surprise that when human beings turn to doing physics, or any other form of thinking, the same organization suggests itself.

The earliest thinking about the physical world was animistic. Physical events were interpreted according to a model drawn from the actions of human beings—as the work of quasi-permanent, conscious agents. Similarly, the model that describes matter as made up of permanent atoms is the projection of our organization of the visual world. The idea of souls is yet another example of this style of thinking, which seeks a substance behind every quasi-permanent form of organization. In this case, it is applied to conscious agents themselves.

To accommodate our increasingly accurate knowledge of the world, we are forced to step away from this familiar and comfortable style of thought. Matter itself is capable of drastic transformations at all levels, down to and including the most basic. What is conserved, in modern physics, is not any particular substance or material but only much more abstract entities such as energy, momentum, and electric charge. The permanent aspects of reality are not particular materials or structures but rather the possible forms of structures, and the rules for their transformation.

Ego and Survival

Imagine you are a scientist who has invented a <u>duplicator</u>, that is, a machine capable of creating an exact duplicate of yourself, including all your thoughts and memories, and projecting it to a remote location. Imagine, too, that malevolent authorities are on your trail. They do not know about the duplicator yet, but they suspect you may have ideas and inventions that they can exploit for purposes of their own, purposes you find completely unacceptable. You know they have methods, unpleasant ones at that, for extracting information from you. They are closing in on you; at any moment, they may burst through your laboratory door.

You realize that you can escape this desperate situation, but at a price. First you will create your duplicate and project it far away. To ensure its survival—your survival—you must put the authorities permanently off the track. You conclude that your best chance for survival lies in suicide, for then the authorities, finding a corpse, will abandon their chase.

We suspect that most people, imagining this story, will find that the suggested strategy, though eminently rational, is emotionally unacceptable. The psychological barriers to suicide are not only, or even primarily, rational. We draw our minds away from situations involving damage to our bodies or, of course, death, as we draw away our fingers from a flame. There are good, biological reasons for these instincts; they help ensure survival and reproductive success and are thereby favored in evolution.

But knowing that our instincts are, at least in some imaginary limiting case, highly irrational is no doubt the first step in taming them.

These thoughts put in a different perspective people's unease upon learning that, for any physical object—be it the earth, a galaxy, an atom, or a human body—the ultimate prospects for survival are poor indeed. What should be most significant to us are not physical ar-tifacts but the meaning they embody. And although duplicators are unlikely to become available anytime soon, it is possible to preserve some of the best and most meaningful parts of our being. Indeed, whenever we create paintings, songs, poems, books, computer pro-grams—or ideas in the minds of children—we do something of this sort.

Inevitability

We have learned that few things are permanent and that those few are rather abstract and intangible. Why the tangible reality around us is what it is, then, becomes a historical question. Here we discuss one important part of this question: Why is the chemistry of the universe, the abundance of different elements, what it is? A richly detailed and satisfying answer emerges: universal chemistry arises from the combination of the original big bang and later little bangs (supernova explosions).

Prelude Four

LEVELS OF EQUILIBRIUM

The idea of conservation, of permanence, is central to our perception of the world. The Swiss psychologist Jean Piaget, in his study of the development of human thought processes, found that the idea of permanent objects is not inborn. Very young children will not search for a vanished toy even when they have seen it being hidden. We gradually learn to translate our perceptions into objects with a useful degree of permanence; at approximately nine months, infants learn that a toy persists even when it cannot be seen.

A long road leads from such primitive concepts of permanence of objects to the refined, abstract conservation laws of modern physics, but the principle involved is the same. Whenever we can say with confidence that some object or quality remains unchanged in given circumstances, we are taking hold of conservation, of uniformity in time.

It should be clear that familiar objects, including our own bodies, are not "conserved" or permanent in the same strict sense as, say, the amount of electric charge in the universe. We can deepen our understanding of the physical world, if we learn to think in terms of levels of conservation.

Under ordinary conditions, most commonplace objects keep their integrity for a long time. I do not ordinarily have to worry that the chair in which I sit is changing or decaying, although in the long run it surely is. The chair is constantly bombarded by molecules of air and often touches other bodies (that's what chairs are for); however, the energy that external air molecules, or molecules at the surface of my Levi's, bring to the collisions is usually not enough to overcome the attractive forces that maintain the structure of the chair. Occasionally, one of the external molecules will have energy well above the average and will chip off a little piece. The cumulative effect of many such rare collisions, given a long enough time, will be to destroy the chair altogether. "Long enough," however, may be very long indeed, and if we are interested in behavior on a much shorter scale of time, the chair may be regarded as permanent.

On the other hand, if the chair is brought in contact with something hot, its claim to permanence becomes more dubious. Frequent hard collisions with more energetic molecules attack its integrity on every side. The chair will burn, or vaporize, in short order.

Remarkably, if we look into the heart of the fire consuming our chair, we find in its very randomness and violence another organizing principle. This is the idea of thermal equilibrium, one of the most useful generalizations in physics. It is a precise version of the thought that in a chaotic situation anything that is not strictly impossible will eventually happen. What is "impossible"? In this context, the only thing that is impossible is to change something that can't change—that is, to violate a conservation law. In thermal equilibrium, the average behavior of any two chunks of matter having the same conserved quantities is the same. When this concept can be applied, it drastically simplifies the task of describing matter, because the number of conserved quantities is generally very small. For example, to describe the properties of a gas in thermal equilibrium, it is sufficient to specify its density and temperature, shorthand ways of talking about the total number of atoms and the total energy, respectively. These are the relevant conserved quantities. It is not necessary to describe the individual positions and velocities of all the vast number of atoms in the gas—these positions and velocities are transitory, changing so rapidly that to measure them at all is a difficult exercise in futility.

Corresponding to the levels of conservation are levels of equilibrium. Generally speaking, the lower the temperature of the system we are interested in, and the less time we allow it to do its thing, the less rigorous is the equilibrium that is established. Fewer and less powerful randomizing collisions have taken place under these conditions, compared with what would have occurred after a longer period at higher temperatures; so more structures can escape their destructive power—more things are effectively "conserved."

Let's haul in our chair again. At ordinary temperatures, and for human time scales, the number of chairs (one) and the orderly arrangement of its molecules into a solid of definite shape are conserved. If the chair material were heated to somewhat higher temperatures, it would vaporize, and neither the arrangement of the molecules nor the number of chairs would any longer make much sense. I could not tell, looking only at this vapor, whether it came from one large chair or two small ones, or anything about the original shape. By contrast, the numbers of each of the different types of molecules that make up the chair material would still be conserved.

At still higher temperatures, the molecules in the vapor dissociate into their constituent atoms; looking at this stuff, I could not tell that it started as wood and paint rather than as an appropriate mixture of pure elements from a chemistry set. And so on . . . A grand dream of physics is that if you heated the stuff up hot enough you could not tell anything at all about it, other than its total mass, and we suspect that at a high enough temperature a pound of chair parts or of gold, Gutenberg Bibles, or positron plasma would all turn out to be the same thing.

The most abstract conservation laws of physics come into their own in describing equilibrium in the most extreme conditions. They are the most rigorous conservation laws, the last to break down. The more extreme the conditions, the fewer the conserved structures we need to worry about. In a deep sense, we understand the interior of the sun better than the interior of the earth, and the early stages of the big bang best of all.

9

Universal Chemistry (or, The Battle Between Entropy and Energy)

Go tell the Spartans, thou that passest by,
That here, obedient to their laws, we lie.
—Simonides of Ceos, Epitaph on Thermopylae (480 B.C.)

The elemental composition of the universe has been deciphered. It represents, in code, the monumental epitaph of a battle that took place ten billion years ago. The battle to decide the chemical fate of the universe was a preliminary skirmish, only minutes long, in a continuing conflict being waged all around us—a conflict whose stakes are the structure of the world. We refer to the war between entropy and energy. Let us introduce the combatants.

Entropy is a pretty word associated with a powerful, but basically simple, idea. Entropy is a mathematical measure of randomness or disorder. (To be precise, the entropy of a system is the logarithm of the number of microscopic states the system can assume. [Don't worry, we won't often be precise.]) A physical system has a large entropy if it can assume many different states in the course of time. For instance, a sample of water vapor at high temperature has much larger entropy than would the same sample cooled down to make ice. The molecules of water vapor are constantly rearranging themselves in space, assuming many different configurations or states, whereas in ice the same molecules are stuck in just one configuration.

For an isolated system, one with no energy or material going in or out, the entropy never decreases. It stays the same, or else gets larger—that's the second law of thermodynamics. Our definition of entropy makes this law easy to understand. By overcoming some constraint, a system may suddenly be able to assume more states than it could before.

For instance, suppose a box of ginger snaps falls off the shelf. Before this event, the cookies are intact. Afterward, they are a disordered mess of fragments and small crumbs. The entropy has increased. We would certainly not expect this reaction to run in reverse. You can't fix your cookies by dropping the box again.

Another, more conventional example is gas enclosed in a container—but let's call it perfume in a little bottle. If the perfume molecules leak out, the entropy increases, because the molecules can arrange themselves in more ways over a larger volume. The second law of thermodynamics predicts that perfume in a little bottle will gradually spread itself throughout a room, but a room with a strong odor of perfume won't unscent itself by concentrating the molecules back into a little bottle. Because there are few ordered states among all the possible states, it is extremely unlikely that a disordered system will become ordered without any prodding.*

If the weapon of entropy is the second law of thermodynamics, how can energy oppose it? Doesn't the second law of thermodynamics guarantee that the universe will become more and more disordered, as every step of its evolution leads to an increase in entropy? Entropy seeks to increase, but the law that energy must be conserved bars the way. Thus, the battle of entropy versus energy is joined. The opponents start with unequal weapons—energy only stays the same, while entropy increases—but any limitations on maximum disorder that are imposed by conservation of energy count as a setback for entropy.

Consider, as an example of an ordered system, the watch on your wrist. To maximize the number of its possible states, it should fall apart into gears and wheels—or, better yet, into molecules, into

---

*We don't want to give you the very wrong impression that it is possible to find important truths about the natural world by playing with words. That is a pernicious idea, which has led many good minds wildly astray. The history of Western philosophy sometimes seems to consist in one fellow's committing this sin and the next fellow's pointing it out.

There are two ways of doing thermodynamics. In one way, you start by talking about readily measurable things like pressure and temperature. Out of these, you construct a combination called entropy that never decreases. It's not easy, however, to show that this measurable "entropy" has a simple meaning, in terms of the possible states of a system. In the other way, you start by defining entropy as we have. Then it is easy to understand why entropy always increases. But in this second approach, it is harder to connect the rather abstract definition of entropy with anything that is easy to measure. Either way, you must make models of matter, and assumptions about how it behaves, before you can connect the conceptually simple with the observable. Something as useful as the second law of thermodynamics, which engineers and chemists use every day, could never emerge from purely semantic arguments.

atoms, into a plasma of elementary particles of all kinds. Why doesn't it?

Large entropy requires many states, but the law of conservation of energy limits the number of states available. So the law of conservation of energy restrains the increase of entropy by holding a system down, limiting it to its lowest energy states. The watch, if left alone, remains an ordered system.

At very low temperatures, very little energy is available, and fewer states are possible. Energy, through its scarcity, becomes the dominant consideration. Matter becomes more ordered; it liquefies or freezes solid. There is ultimately just one state with the lowest-possible energy. When the temperature reaches absolute zero, where just this unique state is possible, the entropy vanishes—a fact sometimes called the third law of thermodynamics.

But try transporting your watch to the atmosphere of the sun; it won't last very long. At high temperatures, there is plenty of energy available, and entropy becomes the dominant factor. Disorder prevails; the watch vaporizes.

The creation of atomic nuclei during the big bang is a contest between entropy and energy. At high temperatures, entropy is dominant. Since there are more ways of arranging protons and neutrons if they are all separated than if some are stuck together, entropy keeps them separate. At low temperatures, the dominant consideration is energy. Although entropy favors separate protons and neutrons, not enough energy is available to break up existing nuclei. In between, interesting things happen, as entropy and energy compete on an even footing.

So much for the simple logic of our program to explain universal chemistry; we now turn to the details of its execution. At exceedingly high temperatures, atomic nuclei as we know them cannot exist. Just as liquid water boils into independent molecules (water vapor) at a hundred degrees Celsius, and as atoms in turn disintegrate into nuclei and electrons (plasma) at temperatures above a few thousand degrees, so nuclei themselves are broken down into protons and neutrons at temperatures exceeding a few billion degrees. In each case, the forces that hold the structures together at low temperatures cannot withstand the more energetic collisions that occur at high temperatures. The stronger the binding forces, the higher the temperature required to overcome them. Our radically conservative working guess, to be judged by its consequences, is that arbitrarily high temperatures occurred early in the history of the universe. Accordingly, there was a time when there were no nuclei, properly speaking, just protons and neutrons. More complicated nuclei could persist only later, when things had cooled down a bit.

Instead, the material that would later become nuclei was a jumble of protons and neutrons, moving too fast to stick together. Although the strong interaction between protons and neutrons is unable to make particles stick at these high temperatures, the weak interaction (whose action does not necessitate any sticking or slowing of the particles affected) could and did play a very important role. Protons and neutrons were bathed in a dense sea of neutrinos and antineutrinos, setting off at rapid-fire rate the dual reactions introduced in chapter 8:

$$n + \nu \rightarrow p + e$$

$$p + \mathcal{V} \rightarrow n + \mathbb{e}$$

These two reactions may look like twins, but the difference in their rates had decisive consequences for the chemistry of the universe. As protons were transformed into neutrons, and vice versa, the relative proportion of protons and neutrons in the world today was determined by the resulting *dynamic equilibrium.*

The idea of dynamic equilibrium is simple but profoundly important. In the early universe, any particular proton (for example) did not survive long before being transformed into a neutron. Of course, new protons were being created from neutrons too. This may sound like a recipe for confusion, but it really describes the conditions for an equilibrium, a balance, based on dynamic processes.

At equilibrium, no individual proton lasts very long. The total fraction of protons in the mixture remains constant, however, because protons are being created out of neutrons just as fast as protons are annihilated to create neutrons. There is equilibrium when the rate of creation of protons (the average rate at which one neutron changes into a proton, times the number of neutrons) is equal to the rate at which protons get destroyed (the average rate at which one proton changes into a neutron, times the number of protons). The important point is that equilibrium is achieved for only one, unique proportion of protons to neutrons. This proportion is simply the ratio of reaction rates. We can go ahead and calculate the reaction rates, or measure them in a laboratory. It is a wonderful thing that by doing so we determine, according to this reasoning, the composition of the early universe. In other words, we can find out just what raw material was available for the next step of creation, the building of the first atomic nuclei.

As the universe cooled down, protons and neutrons could begin to stick together. During a brief period, a building up of nuclei took place. Soon, however, all the available neutrons had been captured in nuclei, and there was not enough energy to create new ones. Nor was there enough energy for the existing nuclei to be disrupted by collisions. Once nuclei neither build up nor break down, element changing stops. The abundances of the different types of nuclei, and thus of the different chemical elements, are "frozen in."

As it turns out, the most important nucleus-building reactions in the early universe cobbled together free protons and neutrons almost in assembly line fashion. Figure 9.1 represents the assembly line in question.

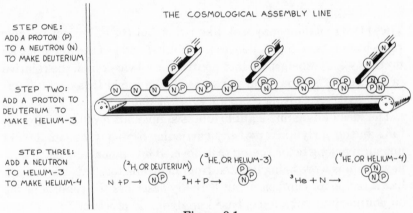

Figure 9.1
The cosmological assembly line.

The net effect of these reactions is that all the available neutrons get funneled into (2P, 2N) nuclei, more conventionally known as ^{4}He.*

*On different occasions we shall use two different systems for naming nuclei. One system is logical: each nucleus is specified by the numbers of protons and neutrons it contains. For example, (1P, 1N) denotes a nucleus containing one proton and one neutron.

The other method of naming elements and nuclei is the standard one, but far less logical. It is based on traditional word names of elements, most of which were invented by chemists before the nature of atoms and nuclei was known. By now, the

At this point, the cosmic synthesis of nuclei stops. The reason that almost no heavier elements arise is simplicity itself: there are no stable nuclei with a total of five or eight protons plus neutrons. It is therefore impossible for a proton or neutron to attach itself to a ^{4}He and impossible for two ^{4}He to stick together. So a good rough description of what happens in the cosmic cooking of nuclei is easy to give: all the available neutrons have been incorporated into ^{4}He nuclei, and the leftover protons remain as hydrogen nuclei.

This may not sound much like the familiar chemistry of Earth, but for the universe as a whole it turns out to be a pretty good approximation. Our radically simple model of nucleosynthesis gives more than just a qualitative prediction. We noted that the primordial ratio of protons to neutrons could be inferred from present-day experiments. Using this ratio, and a little arithmetic, we can calculate just how much helium formed during the big bang.

The mass of the known universe is roughly 75 percent hydrogen, 24 percent helium, and 1 percent minor impurities. This composition is found just about everywhere in the sky. Most stars and interstellar gas, which together make up the bulk of the visible universe, have these proportions. And this ratio of hydrogen to helium very closely matches the calculations based on our very simple model. (The minor impurities—including, of course, ourselves—will be discussed in chapter 11.)

The scientific explanation of universal chemistry, based on a reconstruction of what the universe was like in the first few minutes following the big bang, at temperatures of billions of degrees, is an example of postdiction at its best. Starting from simple principles, simply extrapolated, we discover that the main features of the chemistry of the universe are the inevitable consequences of its early history.

Here, obedient to universal laws, lie the chemical elements of our universe.

---

traditional word names are deeply embedded in the literature; like the crazy scheme of letters on typewriter keyboards, they will probably go on forever.

The traditional name for nuclei (or atoms) containing one proton is hydrogen, or H; the name for nuclei containing two protons is helium, or He; and so forth. These names are based only on the number of protons in the nucleus, because that's what the element's chemical properties depend on. The standard way to add information about the total number of protons plus neutrons in the nucleus is to tack this number onto the element name. For instance, a nucleus of helium 4, abbreviated ^{4}He, contains two protons and two neutrons.

10

Cooking with Gobar
(or, The Energy Strikes Back)

Recycling of chemical energy is a familiar process, and older than you may think. For example, a cow appropriates the chemical energy stored up by plants, by eating them. When the cow, assisted by several cud chewings, multiple stomachs, and millions of symbiotic bacteria, has extracted from its meal all the energy and substance it can, the useless residue is jettisoned. This residue, known in Sanskrit as *gobar,* though useless to the cow, still contains enough energy to have been for centuries the most popular cooking fuel in India.

The recycling of nuclear energy is less familiar, but even older. We live on energy from the sun, energy it generates by nuclear burning. The fuel the sun uses was burned before, during the big bang. Fortunately for us, the burning that occurred in the big bang was incomplete.

It is energy left in the unspent fuel, the hydrogen and the helium, that drives the sun and other stars. The stars recycle nuclei that were the end products of big bang combustion. From the nuclear fuel that the big bang could not fully exploit, they squeeze out the last measure of energy.

Why can stars do better than the big bang? The stars have two things going for them: time and density.

During the big bang, there were only a few minutes when nuclei could form. Very rare processes, or slow ones, played little role. A case in point is the key process from which the sun derives its energy. In this reaction, two protons collide to produce a deuterium nucleus, a neutrino, and a positron. It can be written in several different ways:

$$2 \,^1\mathbf{H} \rightarrow \,^2\mathbf{H} + \text{neutrino} + \text{positron} + \text{energy}$$
$$2 \,(1P, 0N) \rightarrow (1P, 1N) + \text{neutrino} + \text{positron} + \text{energy}$$
$$ⓟ + ⓟ \rightarrow ⓝⓟ + ⓥ + ⓔ + \text{energy}$$

This reaction belongs to the family of weak interactions. All such interactions are rare, but this one is rarer than most, because of the delicate balance of energies involved. In the sun, it takes more than a hundred billion billion collisions between protons to produce a single deuterium. But the sun has plenty of time, and enough protons left over from the big bang to keep this energy-releasing process going for a long time indeed.

When our reaction occurs in the sun, the neutrino escapes, but the positron is absorbed (within a few centimeters, in fact), depositing its energy. The deuterium can then undergo further nuclear reactions, in which still more energy is released, but the production of deuterium is the crucial step. This is because all stable nuclei except for ^{1}H contain neutrons and because, on the other hand, isolated neutrons are unstable. Thanks to the proton-transforming power of the weak interaction, the additional neutrons required to build up heavier nuclei get produced in a stable, bound form—^{2}H.

In fact, the key reaction is just our old friend the decay of the neutron, rearranged and disguised in an interesting way. This appears more clearly if we take away, in our imagination, one proton from each side. Then we are left with a form of neutron decay run backward:

$$ⓟ \rightarrow ⓝ + ⓥ + ⓔ$$

This process, which you might call decay of the proton into a neutron, cannot happen for an isolated proton. Indeed, the proton is lighter than the neutron, so this process is forbidden by the law of conservation of energy—just as a mother cannot give birth to a baby heavier than herself. The extra proton serves as more than just a midwife assisting the proton-to-neutron conversion. It alters the situation entirely by <u>binding</u> to the neutron, to form deuterium. This works because the mass of the bound state (1P, 1N) is not simply the sum of the masses of the proton and the neutron inside—it is decreased by the binding energy. In fact, (1P, 1N) is a little lighter than two separate protons. If it were not, the key reaction that drives the sun would be impossible; it would not be consistent with the law of conservation of energy.

The attractive forces that bind protons and neutrons together are what keep the sun alive. Without an energy source to keep it hot, and thus to generate pressure, the sun would eventually collapse under its own gravity, just as a hot-air balloon deflates when its heat source is turned off. Physicists recognized this as early as the nineteenth century. At that time, it was not known that energy could be generated from nuclear transformations, so it seemed that the sun could not be more than a few million years old. This estimate of the sun's age was much too small for the biologists, who realized that there must have been many millions of years of life on earth for evolution to take place, or for the geologists, who realized that it must have taken many millions of years to deposit the observed thickness of sea floor sediments. Darwin regarded this discrepancy as the most serious difficulty facing his theory and mentioned it at the conclusion of his *Origin of Species.*

This theoretical stumbling block was removed only by the discovery of nuclear energy. It remains, however, a remarkable—and, for humanity, remarkably fortunate—circumstance that the central reaction that drives the sun is so rare. It is only this extraordinary rarity that allows the average proton in the sun to last so long, billions of years, even though it is colliding with other protons millions of times a second. The long lifetime of stars, which makes it possible for biological evolution to work wonders despite its inefficiency and leisurely pace, is an entertaining example of Treiman's theorem.

We talked earlier about how, in the big bang, all the free neutrons eventually got bound into the very stable nucleus of helium 4. A star doesn't start out with free neutrons, but it still ends up funneling its protons into helium 4, by a somewhat more tortuous pathway.

The big bang got stuck on helium 4. It couldn't make larger nuclei, because no stable nuclei result from the most likely collisions—^{4}He with ^{1}H, or ^{4}He with ^{4}He. A star facing the same barrier is pushed to keep on with nuclear burning. For if its energy generation slows down, its core will start to collapse—becoming hotter and denser until the nuclear fires are rekindled. As the material inside the core grows denser and hotter, the star's outer layers simultaneously expand and glow; the next step in stellar evolution, the birth of a "red giant" star like Betelgeuse, is about to begin. Finally, the core gets hot enough and helium itself starts to burn, a process that occurs only when <u>three</u> nuclei collide:

$$3 \ {}^{4}\text{He} \rightarrow {}^{12}\text{C} + \text{energy}$$
$$3 \ (2P, 2N) \rightarrow (6P, 6N) + \text{energy}$$

Such triple collisions are extremely unlikely in the big bang, because by the time helium 4 nuclei have formed, the density is so low that the chance that three nuclei will be in the same place at the same time is negligibly small. Even in stars, which are much denser, triple collisions are pretty rare. That is why helium-burning red giants can get energy from helium 4 for hundreds of millions of years before exhausting their supply.

Looking at stars as recycling centers for energy left over from the big bang (a sort of ecological perspective), though unusual, is helpful and revealing. One thing it makes us appreciate is just how delicate the relationships are that make our world the way it is. If the forces holding protons and neutrons together were just a little bit different, so that there could be stable nuclei with a total of five or eight of these particles, the big bang would have burned the nuclear fuel almost to completion, leaving nothing for stars to run on. If the weak interaction were a little stronger, the sun would evolve much more quickly into a red giant, with catastrophic consequences for life on Earth. And so on. . . .

The delicate balances that make nuclear energy available to

stars—but not too quickly—are remote and seemingly accidental consequences of certain fundamental laws of interactions. Because tiny differences in these fundamental laws could produce such enormous consequences, it is impossible, as a practical matter, to compute the character of the natural world by starting from the fundamental laws. We have to let nature do some of the more delicate calculations in the middle and take her word for the results.

11

Explosions and Fluorescence
(or, Entropy's Revenge)

Things fall apart; the centre cannot hold;
Mere anarchy is loosed upon the world. . . .
—W. B. Yeats, "The Second Coming" (1921)

The story is so far only half-told. After all, we can hardly be satisfied with a description that leaves our planet—and ourselves—quite unaccounted for. The big bang produced hydrogen, some helium, and very, very little else. Where do the other, heavier elements come from? There is a strong scientific case that the heavier elements are produced in stars and liberated in great entropy-enhancing explosions.

That brings us to the story of this chapter: how stars, following the logic of their development, inevitably turn themselves into stupendous powder kegs.

We have already followed the history of a star from the early, hydrogen-burning to the helium-burning, red giant phase of its existence. Hydrogen and helium could be compared to the slow-burning chemical fuels firewood and charcoal—or *gobar*. And just as the charcoal left over from burning wood can be burned again, so the helium produced by burning hydrogen is recycled within the star to yield more energy.

When helium is exhausted, the process of collapse, heating, and rekindling the ashes begins again. The ashes of earlier burning stages become the fuels of later ones, as stars are forced to attempt to squeeze every last possible drop of energy from their own innards.

An important difference separates hydrogen and helium burning from the later burning stages. Both hydrogen and helium burning are necessarily slow. Hydrogen burning is slow because its key step involves a rare type of weak interaction; helium burning is slow because it requires rare triple collisions. These facts allow stars to burn for a very long time without using up all their fuel.

Once helium burning has occurred, however, the next possible reaction—*carbon burning*—is not necessarily slow:

$$2 \; {}^{12}\text{C} \rightarrow {}^{24}\text{Mg} + \gamma$$
$$2 \; (6\text{P}, 6\text{N}) \rightarrow (12\text{P}, 12\text{N}) + \gamma$$

Lots of energy is liberated in this reaction, some of it in the form of a high-energy photon, or gamma ray.

This reaction involves a simple rearrangement of protons and neutrons, what we call a strong as opposed to a weak interaction. And, of course, it requires only an ordinary collision, not one of those rare triple collisions. So if the temperature is high enough, carbon burning will proceed very rapidly.

In more detail, this is the situation: the carbon nuclei repel one another when they are far away, by ordinary electric forces, because both have positive electric charge. However, once they can manage to get close enough together that they almost touch, much stronger attractive forces will come into play and cause them to bind together, liberating energy. (This simple plot line has sold many a romantic novel.)

Long-range repulsive forces and more powerful but short-range attractive forces are competing to determine the fate of the carbon nuclei. As a result, the reaction's rate depends sharply on the energy of the initial nuclei or, in other words, on how fast they're moving. Only fast-moving, high-energy nuclei will manage to collide, smashing past the barrier of repulsion to fuse at last.

At high temperatures, the carbon nuclei will routinely be moving fast enough to get together, and the burning will go like gangbusters. It is even a potentially explosive situation, because the energy liberated in the burning keeps the temperature rising. But at low temperatures, <u>nothing</u> happens. In this way, a star builds up a store of material, which only awaits a spark—a sudden rise in temperature—to detonate it.

If the temperature is maintained just right, stars can have a relatively brief period—a few thousands of years—of more or less well controlled carbon burning. Carbon burning results in magnesium. Our discussion of how carbon burns applies word for word to magnesium, which is explosive for the same kind of reason. Taking a cross section of a highly evolved star would reveal a system of many layers. The inner layers have been subjected to the largest pressures, thereby forced to the highest temperatures, and burned the furthest; the outermost layers, by contrast, have not burned at all. Thus, as we proceed from the outside in, there will be an outermost layer with the initial mix of hydrogen and helium, a layer of mostly helium, a layer of carbon, a layer of magnesium, and so on. We have compared the nuclear burning of hydrogen and helium to the slow combustion of chemical fuels like *gobar,* wood, or charcoal. Carbon and its successors are more like TNT, which will explode if given the right spark.

So we arrive at the picture of a star, in the late stages of its evolution, having gone through several stages of nuclear burning and now composed mostly of carbon nuclei and other explosive material. Eventually, the possibilities for generating energy by nuclear transformations run out. The protons and neutrons have been arranged to take maximum advantage of their attractions for one another; no further energy can be squeezed from them.

When the center of a star reaches this state, as it eventually must, its energy source has failed. It therefore starts to cool, and the internal pressure drops. But it is just this pressure that supported the structure of the star and kept the great mass of the overlying layers from "falling in." As it did when it switched from one fuel to another, the star enters a transition state of collapse, compression, and heating.

This time, however, things go further. There is no new pressure source for the core; the compression and heating continue. Explosive material, pulled into this inferno, ignites and detonates. A supernova is born. The awesome power of an H-bomb derives from the fusion of a few grams of material. Here, we contemplate the fusion of a trillion trillion tons.

Supernova explosions catapult the material of entire highly evolved stars into interstellar space. The ejected material is very different from the mix of hydrogen and helium that went into making the star in the first place. It is rich in heavier nuclei. Multiple cycles of nuclear burning, and further nuclear reactions occurring during the explosion itself, have added heavier elements to the mix. And this is how universal chemistry is enriched with "impurities."

The result of a star's attempt to obey the full command of energy and to reach the most perfectly bound state of protons and neutrons, with the lowest possible energy, ends in an explosion. It is a victory for entropy, for disorder.*

We have presented our account of stellar evolution, and creation of nuclei, in narrative style. The arguments, we trust, seem logical, but to be scientific we must seek more. "It's a nice story—but are you sure it's true?" the skeptic is entitled to ask. To satisfy this proper doubt, we must search for observable consequences of our story, for relics to postdict. But this search is not undertaken merely to satisfy the skeptic. It is delightful in itself when we are able to interpret features of the present as signs confirming our understanding of the past.

There is an intricately constructed web of evidence that heavy elements (that is, everything beyond hydrogen and helium) originate mainly in supernova explosions. We will follow only a few representative strands.

The most fundamental piece of evidence is simply that the <u>same pattern</u> of abundances of different heavy nuclei has been observed in meteors, in many stars, and in the interstellar medium. This is consistent with the idea that stellar explosions, which occur throughout the galaxy, are responsible for producing the pattern.

The idea that stellar explosions are the main source of heavy nuclei powerfully elucidates many details in the pattern of universal chemistry. For one thing, the nuclei produced in the various cycles

---

*The second law of thermodynamics was never repealed for stars. While it is true that the nuclei are being cooked into a more ordered state, that's not all there is to it. Stars shine, emitting light into the surrounding spaces. This light represents energy in a highly disordered form; viewed over short periods of time, it flickers like mad. It therefore carries a lot of entropy. The <u>total</u> entropy, of nuclei plus light, increases during the cooking. But the ordering influence of energy wins a partial victory and governs the evolution of the <u>matter</u> in stars.

of stellar burning—carbon, magnesium, and so on—are observed to be the most common of the heavy nuclei. (Remember that in this context everything beyond hydrogen and helium is considered heavy.) This is just what we expect if stellar explosions are the source of heavy nuclei, because the material of the exploding star consists precisely of the products of stellar burning. Other subtle regularities in the table of abundances can be explained by calculating out the nuclear reactions expected to occur in stellar explosions.

Some direct observations also support this picture. A few supernovae have been located shortly after their eruption. Astronomers then can study the abundant light emitted in the aftermath of the explosion, and check directly that heavy elements are being ejected. Specifically, they look for the spectral lines of heavy elements in the light of supernova explosions. A mixture rich in heavy elements is found, as our models would lead us to expect. Recently, astronomers have even discovered, among the debris of supernova explosions, shells of different chemical composition—just the sort of shells stars are expected to form during different stages of nuclear burning. By exploding, the star has in effect exposed its interior to inspection. These recent discoveries dramatically confirm theories about the later stages of stellar burning and, of course, very much encourage the idea that heavy elements originate in supernovae.

Yet one more piece of evidence: the oldest stars, when compared with younger stars and with the interstellar medium, are deficient in heavy nuclei. This fact would be very difficult to understand if the heavy nuclei were primordial, or antedated the formation of stars. On the other hand, it is entirely consistent with the idea that the heavy elements were produced in stellar explosions. Then the very oldest stars would have condensed from material that had not yet been enriched by supernovae, whereas the material from which later generations of stars formed—and, of course, the interstellar medium we now observe—is enriched by the ejecta of earlier generations of exploding stars.

Altogether, there is a circumstantial but compelling case that the heavy elements are created in stars and ejected in stellar explosions.

Let us now apply these ideas to a case that, although insignificant on the cosmic scale, is of special interest, the case of Earth. The pattern of element abundances on Earth is quite different from the pattern for the universe as a whole. For one thing, hydrogen and helium, which predominate in the universe, are minor components of Earth. The explanation of Earth's atypical composition is surely to be found in the history of its formation. Hydrogen and helium gas solidify only at extremely low temperatures and high pressures. Although the history of how Earth formed is not completely understood, it is hard to imagine how these gases could have solidified in the neighborhood of proto-Earth. On the contrary, it is very natural to suppose they remained as gases. As gases go, both hydrogen and helium are extremely light. They would escape easily from the atmosphere of proto-Earth, which was hot and not fully condensed, so that the atoms had plenty of energy to escape from a less powerful gravity. The present Earth, according to this reconstruction of events, is the residue of a much more massive primordial cloud, whose volatile elements—including the hydrogen and helium that made up most of its mass—leaked away or were boiled off.

Some indirect evidence for this idea is provided by the outer planets. The giant planets—Jupiter, Saturn, Uranus, and Neptune—are mostly hydrogen and helium. They are giants simply because when they condensed (at low temperatures compared with those of the more sun-warmed proto-Earth), the volatile gases did not boil off. These planets have satellites, much smaller than themselves, with more Earth-like composition. The proto-satellites must have lost their coating of hydrogen and helium to the powerful tides exerted by the central planet.

Although various chemical processes have altered the elemental composition of Earth from the universal one, one important part of the record is preserved. Different isotopes of the same element do not get separated in any reasonable chemical process, including just about any conceivable process in the formation of proto-Earth. So we should expect the relative abundance of different isotopes of the same element to be universal, and in particular the same for Earth material and meteor material. With a few interesting but understandable exceptions, this is found to be so. The Earth, we conclude, consists of the same material—stardust, from stellar explosions—as heavy elements everywhere else.

Everyone is familiar with toys that glow in the dark. They are made with materials that capture the energy of light and hang on to some portion of it for a long time, emitting it very slowly. The normally rapid process of emission somehow gets "hung up." Such materials are said to be *fluorescent.* Whiteners, in fabrics and detergents, employ fluorescent substances that capture high-energy ultraviolet light (emitted copiously by the sun, but invisible to us) and re-emit it in dribbles, as visible light. You can actually see your clothes glow in a sufficiently dark room, especially if they have just come in from the sun. Fluorescence in the traditional sense is associated with visible light or its near relatives; at a microscopic level, it is due to rearrangements of electrons in atoms and molecules.

There is a more energetic form of fluorescence, associated with rearrangements in the interior of nuclei. Nuclei, after capturing energy that rearranges them, may get hung up in configurations that take a long time to unravel. The resulting nuclear "fluorescence" is called *natural radioactivity.* In natural radioactivity, the energy is released not as ordinary light but in the form of gamma rays, alpha particles, and other exotica. The principle is exactly the same, however. Natural radioactivity is the slow release of part of the energy captured by nuclei in stellar explosions, just as ordinary fluorescence is the slow release by atoms of part of the energy they capture from light.

In this sense, all of Earth glows in the dark. The energy released by natural radioactivity, the lingering fluorescence of stellar explosions, keeps Earth dynamic. It melts the core and keeps it flowing, and heats the crust and mantle, with consequences ranging from the generation of earth's magnetic field to earthquakes and the motion of continents. By contrast, Luna and Mars, because they are smaller, derive less energy from radioactive decays. Geologically, they are dead.

In this chapter and the two preceding ones, we have traced the nuclear history of the world, the history of its elemental composition. Our reconstruction of this history has allowed us to make quantitative predictions (for instance, the relative abundance of hydrogen and helium) and to understand some surprising details (for instance, the differences in composition between the old stars and those formed recently). Such achievements are the hallmarks of a correct and fruitful theory. We can be confident, therefore, of a most remarkable conclusion: the chemical composition of the world is the inevitable product of its evolution according to known physical laws.

are said to be *fluorescent*. Whiteners, in fabrics and detergents, employ fluorescent substances that capture high-energy ultraviolet light (emitted copiously by the sun, but invisible to us) and re-emit it in dribbles, as visible light. You can actually see your clothes glow in a sufficiently dark room, especially if they have just come in from the sun. Fluorescence in the traditional sense is associated with visible light or its near relatives; at a microscopic level, it is due to rearrangements of electrons in atoms and molecules.

There is a more energetic form of fluorescence, associated with rearrangements in the interior of nuclei. Nuclei, after capturing energy that rearranges them, may get hung up in configurations that take a long time to unravel. The resulting nuclear "fluorescence" is called *natural radioactivity*. In natural radioactivity, the energy is released not as ordinary light but in the form of gamma rays, alpha particles, and other exotica. The principle is exactly the same, however. Natural radioactivity is the slow release of part of the energy captured by nuclei in stellar explosions, just as ordinary fluorescence is the slow release by atoms of part of the energy they capture from light.

In this sense, all of Earth glows in the dark. The energy released by natural radioactivity, the lingering fluorescence of stellar explosions, keeps Earth dynamic. It melts the core and keeps it flowing, and heats the crust and mantle, with consequences ranging from the generation of earth's magnetic field to earthquakes and the motion of continents. By contrast, Luna and Mars, because they are smaller, derive less energy from radioactive decays. Geologically, they are dead.

In this chapter and the two preceding ones, we have traced the nuclear history of the world, the history of its elemental composition. Our reconstruction of this history has allowed us to make quantitative predictions (for instance, the relative abundance of hydrogen and helium) and to understand some surprising details (for instance, the differences in composition between the old stars and those formed recently). Such achievements are the hallmarks of a correct and fruitful theory. We can be confident, therefore, of a most remarkable conclusion: the chemical composition of the world is the inevitable product of its evolution according to known physical laws.

Quantal Reality

The reality of the microworld, quantal reality, demands the deepest revision and the most breathtaking expansion of man's concept of the world ever made. Quantum mechanics is the strangest, and the most successful, of scientific theories. Here we begin by looking at simple examples of the apparently contradictory behaviors that nature exhibits in the microworld, then see how these behaviors are reconciled in quantal reality.

IN THE FIRST CIRCLE

Near the entrance to Dante's Hell lies the First Circle, also known as Limbo. It is populated by the virtuous heathen, those who failed to accept the Christian faith and yet refrained from serious sin. Dante found there not only Plato and Aristotle, born too soon to know the faith, but also Saladin and Averroës, Moslem and Jew respectively, who were born during the Christian era and lived virtuous lives outside the church.

We suspect that many, perhaps most, modern scientists will find their way to the First Circle. If the place is as Dante described it, they will find it congenial, an elite retirement community where the shades of the dead can carry on old arguments, amid a distinguished company. According to Dante, the souls in Limbo suffer because they are aware of celestial regions, inhabited by the blessed, which they can never reach. We wonder, though, whether some might actually prefer to say closer, both figuratively and literally, to what they came in life to regard as the "real world."

In the First Circle, where there is plenty of time and everyone gets acquainted, a remarkable friendship sprang up among a philosopher, a mystic, a musician, and a physicist. During one encounter, they set out to explore the ultimate nature of the reality that lies behind appearance.

The philosopher held, like Plato, that our senses give us only a very partial and often misleading picture of the world. "Let me show in a figure," he said, "how far our nature is enlightened or unenlightened: Behold! human beings living in an underground den, which has a mouth open toward the light and reaching all along the den;

here they have been from their childhood, and have their legs and necks chained so that they cannot move, and can only see before them, being prevented by the chains from turning round their heads. Above and behind them a fire is blazing at a distance."

"This is a strange image, and these are strange people,' said the musician.

"Like ourselves," replied the philosopher, "they see only their own shadows, or the shadows of one another, which the fire throws on the opposite wall of the cave."

The mystic was not unhappy with this, as far as it went, but had something to add. "There are an infinity of worlds," he said, "for the extent of Being cannot be limited. And among these worlds there must be many like our own, or differing only in infinitesimal details. From these infinitesimal differences, great divergences arise in time. In an infinity of worlds, Helen of Troy has a wart at the end of her nose; in another infinity, Nero is bitten by a mosquito and dies of malaria before Rome burns. Each of these worlds, though identical in every other respect to ours, develops an entirely different history. There are worlds in which I inhabit the Second Circle,* and others in which I sit with the blessed. Yet such is the inexhaustible infinity of worlds that there remain infinitely many close to ours in every respect at this very moment, and so it will be in the future."

"But where are these worlds? Can we visit them or communicate with their inhabitants?" asked the musician.

"And if two worlds can be bridged, are they not in reality two segments of a single world?" asked the philosopher.

The mystic was silent for a time. He was not used to answering questions, since he seldom asked them of himself. His style of thought was to visualize, rather than to analyze. Finally, he said to the musician, "Imagine a labyrinth with very many paths. Two men begin at the same point but choose different directions at one fork. The labyrinth is so intricate that though they wander forever they will probably never again encounter one another. They inhabit the same space, and their paths may even cross. But the chance that they meet again, that they come to the same place at the same time, is vanishingly small. Just so the different worlds." To the philosopher, he said, "You are too fond of words, perhaps because you are so clever with them. Two worlds that can hardly communicate with

*The Second Circle houses the lascivious.

one another can be called a single world if you insist, but this does not change the reality. It is as if you called our two friends wandering in their labyrinth a team of explorers. If they never meet to compare their experiences, they are not a team in fact."

"It delights me," said the musician, "that I can easily understand what each of you says in my own terms. Can there be realities beyond common awareness? I have no doubt of it. I need only imagine myself visiting a village of deaf people. And surely the music that Bach hears is a very different and much richer thing than what an untrained person hears, although it may be the same sounds. Bach perceives harmonies and conceives possibilities for variation of which the tyro has absolutely no conception." Turning to the mystic, he added, "As you spoke of branching parallel worlds, I imagined voices starting in unison, then developing different variations on the original theme, at first barely discernible, but eventually becoming independent elements in a full-scale fugue."

"As for myself, I must admit to a faith I can in no way justify, except that I find it beautiful to imagine. I think of the world as a musical instrument, a sounding board on which some gigantic melody is being played out. Everything we sense is a vibration of this universal sounding board, its coming to be and passing away the passage of a note in the unfolding melody."

"This is remarkable!" exclaimed the physicist, and his friends prepared themselves to hear some of his usual gibes. "I don't know how you guys did it," he said, shaking his head. "I'm amazed. You've all got it basically right. Let me tell you about quantum mechanics . . ."

12

Light as Waves

I consider that I understand an equation when I can predict the properties of its solutions, without actually solving it.

—*P. A. M. Dirac*

In preparing to enter the subatomic microworld of quantum physics, you must proceed with as much caution as if you were lacing up your first ice skates. The commonsense reactions built up over a lifetime of ordinary experience will not serve you well in the slippery realm before you.

Quantum mechanics can seem outrageous. For example, we find that in the subatomic microworld any path from causes to effects is constantly blurred by uncertainty and jogged from side to side by randomness and chance. Can this be what underlies the ordered and inevitable universe revealed by classical physics? Stranger still, our universe itself seems continually to branch in all directions, at each moment giving birth to an almost unimaginable infinity of coexisting worlds. In coming to terms with quantum physics, we are forced to expand quite drastically our vision of physical reality.

We cannot, of course, make our minds blank slates. Nor should we. To begin to "understand" the microworld, in Dirac's sense, we must approach its strangeness through analogy and metaphor, linking it to things we comprehend more clearly. In the end, we skaters learn to exploit leg muscles designed and trained for quite different purposes to speed us along the ice.

The concepts of quantum mechanics, though treated here in metaphorical rather than mathematical language, are definite, not hazy. The strength, and much of the fascination, of these strange ideas is that they spring from concrete—and not even particularly complicated—experimental facts about the behavior of light and matter. No fantasy world is more fantastic, and no revealed "truth" more awesome, than the scientific reality of our own universe, inferred from humble facts that anyone can observe and understand.

We shall root the discussion of quantal reality in two ideas that describe the behavior of light in different circumstances: *waves* and *particles* (or, as we will often say, lumps). The wavy behavior of light will be discussed in this chapter; its lumpy behavior, in the next. Each behavior, separately, is easy to understand. What is very difficult is to see how they avoid contradicting one another—how one thing can at the same time be both wave and lump.

Waves and lumps represent two very different ways of getting energy and information from one place to another. If we are receiving energy or information, it seems a straightforward task to tell whether it arrives in a lump or by way of a wave. Particles and waves behave quite differently.

For one thing, waves will spread out and bend around an obstacle, a phenomenon known as <u>diffraction</u>. Because ocean waves diffract around people, even a summer crowd of swimmers creates no breaks in the wavelets rippling onto the sand. The path of a particle, on the other hand, does not bend around obstacles. A machine-gun bullet fired close to the edge of a wall will either hit the wall or miss it. If it hits the wall, it stops; if it misses, it goes straight: no diffraction. This suggests a simple way to distinguish waves from particles: hide your detector behind a wall. If the energy still gets in, you have a diffracted wave; if the wall blocks it, you have a straight-shooting particle.

Another way particles differ from waves is by their *discreteness,* or lumpiness. No one has ever seen half an electron. Either you have an electron, bearing its full freight of mass and electric charge, or you have none. Any properly working electron detector will always count whole numbers of electrons, never fractions. In contrast, if you set up a detector at the beach to measure the energy in ocean waves, you can stick it into the water in the middle of a wave and pull it out again whenever you like. You can measure the energy given off by three and a quarter waves in precisely the same way as you can measure the energy of a single wave. Waves aren't discrete; they're continuous.

The behaviors of waves and particles seem different to the point of incompatibility. Yet light exhibits both diffraction and discreteness. From this paradox, unavoidable because it is in the observations themselves, the strangeness of quantal reality will flow.

Ocean waves are familiar but complex, in many ways deviating far from the mathematical ideal of wave behavior. We want now to take a closer look at a simpler wave system.

Imagine a rapidly pulsating sphere—a sphere that expands and contracts rhythmically, many times a second. Such a sphere generates a monotonous humming tone as it sends a sound wave pulsing through the air. We want to understand what is going on in this simplest-possible case of wave generation, so we can work our way up to trickier cases. When the sphere is expanding, it hits some nearby air molecules and pushes them outward into the adjacent region. This region, however, is already occupied by a normal density of other air molecules. The resulting overcrowding creates a high-density, high-pressure layer around the sphere. When the sphere contracts, just the opposite occurs. The shrinking sphere is vacating a space momentarily free of air, and a low-density, low-pressure layer results.

Our sphere strikes only molecules nearby. The resulting disturbance, however, takes on a life of its own. The molecules that hit the sphere collide with a second layer of molecules and transfer some of their momentum. The second layer strikes a third, and so on.

Imagine the traveling sound waves far from the sphere. Each high-density, high-pressure layer tends to squeeze out its excess air molecules. As they flow into a previously average region, that region in turn attains high density and high pressure. And so, in a self-perpetuating cycle, the disturbance expands outward. In this way, news of the sphere's pulsation travels across long distances.

Metaphorically speaking, our pulsating sphere gives marching orders to the air immediately outside. The air transmits these orders onward, at the speed of sound. Shells of compressed and rarefied air spring up in obedience. The air at any point is compressed, or rarefied, depending on what the sphere was doing some time ago. The time delay is how long it takes to transmit the orders—the distance divided by the speed of sound.

Thus rhythmic pulsations of our sphere become rhythmic changes in the density of the surrounding air. We interpret such changes, when they transmit their energy from the air to our eardrum, as sound.

Having discussed the result of one pulsating sphere, let us boldly proceed to consider two identical pulsating spheres. This is actually no joke, because some interesting new features appear when we think carefully about this case. Each sphere sends out its marching orders, as before—but these orders may disagree. How does the air attempt to serve two masters?

Consider the air at a point equidistant from the two spheres. It takes its instructions, on whether to compress or rarefy, on the basis of what the spheres were doing some time ago. Since the distances to the two spheres are the same, the time delay is the same. If the spheres vibrate together, the air is always getting identical instructions from both. It compresses and rarefies more vigorously—with twice the amplitude, we say.

Consider, next, air at a point a little farther from one sphere than from the other. The delay times are no longer equal, and the point will receive different orders from each sphere. In the special case where the orders arriving at some point contradict each other directly, the density serves both as best it can, by not changing at all. The result will be points of quiescence—technically known as *nodes*—where the air's density varies not at all, and no sound is heard. Note the paradox here: either sphere alone creates a sound wave at this point; two spheres together add up to no sound there at all. Two sources can add up to give less than one. This is the essence of *destructive interference*.

(When two sources are giving the same instruction, the resulting vibration bears not just twice but four times the energy. This phenomenon, oxymoronically known as *constructive interference*, may seem puzzling. Just imagine trying to jam all the riders from two full subway cars into the volume of a single car, and you will appreciate that this can take much more than the sum of the energies to pack each car separately.)

Space is filled with an intricate pattern of enhanced vibration and canceling vibration. To visualize this interference pattern, it helps to simplify the geometry a bit. A convenient arrangement (both for sound and, as we shall see, for light) is shown in figure 12.1.

Figure 12.1 should be viewed as a snapshot of interference between waves emitted from two identical sources. As time goes on, the patterns march out from the source. The identical sources here are two narrow slits in a screen, each transmitting waves from the same primary source. The curved solid lines represent the high-density regions (for sound) or, in general, wave crests. The total

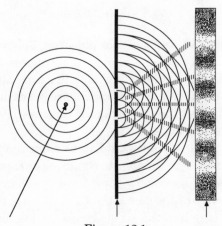

Figure 12.1

Primary source   Secondary source   Detection screen

motion will be particularly vigorous where two such lines intersect.

Suppose the wave sources in question are sources of light. If light consists of waves, constructive and destructive interference at the detection screen will create an alternating series of bright and dark regions, as shown. That we do see interference for light strongly hints, of course, that it consists of waves.

Figure 12.2 depicts another way of looking at interference measurements. Here, the shaded area represents the intensity of disturbances (loudness of sound or brightness of light) arriving at the detection screen.

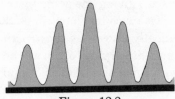

Figure 12.2

The peculiar addition law for interfering sources is that, when the energies of waves are added, one plus one is not simply two. Either slit alone gives (essentially) uniform intensity. But when both are open, the results do not simply add. Exaggerating slightly (replacing, in Figure 12.3, the dotted by the solid lines), we find that the essence of the matter is a strange new rule of addition. One plus one is sometimes zero and sometimes four; two is only the average. It is a most peculiar sort of addition.

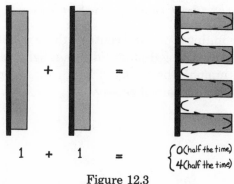

$$1 \quad + \quad 1 \quad = \quad \begin{cases} 0 \text{ (half the time)} \\ 4 \text{ (half the time)} \end{cases}$$

Figure 12.3

The wavelengths of sound are on a human scale—about the length of a piano or guitar string. The wavelengths of light, on the other hand, are only a few hundred times the size of an atom. Just as the wavelengths of visible light are on a smaller scale than those of sound, so are all its wavelike behaviors. For example, diffraction is far less pronounced for light than for sound—that's why we can hear around a corner but not see around one.

Nevertheless, in a few everyday examples the interference effects for light show themselves. One that can be exquisitely varied and beautiful is the appearance of a soap bubble. The colors of a soap bubble result from interference between light reflecting off the top of the film of soap and light reflecting off the bottom. Because the soap film's thickness is comparable to a wavelength of visible light, constructive interference creates bright spots of the appropriate color. Of course, the thickness of the film varies, and so different colors show up in different places. Similar effects occur for oil slicks. The striking tail colors of the male peacock are also the result of interference, this time between light reflected off cleverly spaced layers of feathers.

Although everyday examples of the interference of light waves are few, the scientific examples are many. The beautiful, highly developed art of describing wave behavior in mathematical terms can predict light's behavior in elegant detail. For example, we can predict the patterns formed by many interfering sources (instead of just two) or by sources arranged in complicated ways. The wave theory of light is also used to describe the details of the way lenses expand or focus a light beam, and hence in designing eyeglasses, telescopes, and microscopes. The idea that different colors arise from light waves of different wavelengths has been tested in exhaustive detail.

Around the beginning of our own century, there were only a very few scientific voices dissenting from the idea that light travels as a wave. Given all its triumphs, the theory of light as waves must be said to capture an important part of reality.

13

Light as Lumps

*Query 29. Are not the Rays of Light very small
Bodies emitted from shining Substances?*
—I. Newton, *Opticks* (1706)

In spite of the great successes of the light wave, let's take a second look at Newton's idea. If light at its most basic level were made up of lumps instead of continuous waves, what would we observe?

Consider a light bulb. It certainly seems to send out light continuously and in all directions, much as the pulsating sphere sends out sound waves. We don't expect to see water molecules while gazing at a river, however; and if individual light particles were crudely apparent, we would already know about them. To look for lumps, we must turn the light way, way down. Cover it with a sheet of muslin, then with another, and another. As its glow get softer and softer, what do we ultimately see?

If light came out as radial waves, we would see a continuous flow of feeble radiation, simultaneously getting paler and paler in all directions. All our instruments would report the same amount of light coming out of the bulb, until it became so pale they could no longer detect anything at all. This, however, is <u>not</u> what sensitive light detectors record.

The glow of a weak source is not continuous. It appears instead as a series of flashes. What happens as the source gets weaker is that the flashes get rarer, while the size of each individual flash remains the same. If we hook up our detectors so that their readout is a sound, we shall hear not a steady hum but click . . . click . . . click. . . . <u>Light comes in lumps</u>.

Furthermore, the flashes don't spread out spherically like waves. Each detector emits its own, random pattern of clicks. They never sound together, because each lump of light gets into only one detector. Lumps of light go off in definite directions, like particles.

If light comes in lumps, interference suddenly grows very paradoxical. As we discussed in the last chapter, interference is a wave phenomenon. It arises when two disturbances send out "marching orders" through all space, which sometimes agree and sometimes don't. But lumps don't fill all space; they go in particular directions.

At a still more basic level, the very idea of interference just doesn't seem to make much sense for lumps. When there's a conflict of marching orders, waves can compromise by adjusting their amplitude. The basic fact about lumps of light, however, is that each one is an irreducible *quantum*. (This word, from the same Latin root as *quantity*, just means a basic, elemental unit.) You can have one lump or two, but not half or a quarter of a lump. The addition law in figure 12.3 (the dotted line) just can't be implemented in lumps.

If you can't stand the tension, please turn immediately to the next chapter, where this paradox is resolved. For cooler heads, we will now mention a few other interesting points about light as lumps or, more formally, *photons*.

We have spoken of pure colors as being light made up of a single, specific wavelength. We could equally well state that such light waves show a single, specific *frequency*. This works because the frequency of a wave measures how frequently the light wave goes through a complete cycle: the speed of light divided by the length of a single wave gives us its frequency, or the number of waves per unit of time.

How do we reconcile colors with photons? Experiments also show that each pure color (or frequency, or wavelength) of light is made of photons, with a particular energy. Binding together wave and particle ideas, one finds that the energy in a photon is proportional to the (wave) frequency of the color it represents.

Einstein proposed the concept of photons in 1905. He also proposed that each photon's energy is given by just the frequency multiplied by a universal constant known as *Planck's constant*. It was ironic, to say the least, that Einstein should have introduced a particle model of light in terms of frequency, which is an attribute of waves and not of particles. His proposal rested on very indirect evidence and seemed to fly in the face of everything physicists had learned about light in the preceding hundred years. Even eight years later, a distinguished panel of physicists (including Planck himself), who were recommending Einstein's appointment to the Prussian Academy of Science, felt compelled to apologize for this aberration on his part:

> Summing up, we may say that there is hardly one among the great problems, in which modern physics is so rich, to which Einstein has not made an important contribution. That he may sometimes have missed the target in his speculations, as, for example, in his hypothesis of light quanta, cannot really be held too much against him, for it is not possible to introduce fundamentally new ideas, even in the most exact sciences, without occasionally taking a risk.

It seemed incredible to these experts that a serious scientist would dare to tamper with the brilliant successes of the wave theory of light. In science, however, the ultimate judges are not experts but experiments. In his original paper, Einstein mentioned a crucial experimental test of his ideas, which was carried out ten years later (straining the technology of the time to its limit).

His idea concerned the *photoelectric effect,* an effect routinely used today, among other things, for opening doors when a light beam is interrupted. The basic underlying phenomenon is that you can knock electrons out of a metal by shining light on it. Einstein pointed out that by looking at the velocity of the ejected electrons, experimenters could test his photon theory of light.

If light of a definite frequency transferred a definite amount of energy, as in his photon theory, the velocity of the escaping electrons should depend on the frequency of the light, not on its intensity. If light comes in lumps, more intense light just means more lumps. So more electrons will be knocked out of the metal, but each one will get the identical kick.

A wave picture gives just the opposite expectation. If the energy in the light is spread evenly and continuously through space for the electrons to pick up, then increasing the intensity of the light should make the electrons move out more energetically. Also, there is no reason to think that every electron gets the same kick; the kick received will depend on the precise strength of the light wave penetrating the metal, which varies with depth.

The experiments showed that Einstein was right. The energy of light is transferred in lumps.

In our discussion so far, we have tacitly assumed that "light" means "visible light." But visible light is only one, restricted case of the more general concept of electromagnetic radiation. Visible light is radiation with wavelengths between 4×10^{-5} and 7×10^{-5} centimeters—or, equivalently, frequencies between 4×10^{-14} and 8×10^{-14} cycles per second. It corresponds to just one octave amid the infinite range of possible radiation.

Light of longer wavelengths ranges from the infrared up to microwaves and radio waves. The longer the wavelength, the less energy in each lump, or photon. For radio waves, discreteness is well hidden, because it takes many lumps to yield a noticeable amount of energy—just as water seems the epitome of smoothness because it is made up of enormous numbers of tiny molecules, or as a news photo made up of many black dots can look gently shaded.

Going in the other direction, toward shorter wavelengths, we pass from ultraviolet light to X rays and to gamma rays. Gamma rays are detected with Geiger counters. Our little clicking experiment with phototubes is a fraternal twin of the click of Geiger counters when hit by gamma rays. But whereas the particle character of visible light is subtle and ambiguous, gamma rays appear "obviously" to be particles, because each gamma-ray photon packs an easily noticeable punch.

We come to appreciate that in solving the paradox of how visible light can be both waves and lumps, we will not only relieve an intellectual discomfort but also unify the description of what appear to be hopelessly diverse parts of nature. Gamma rays are "obviously" lumps, and radio waves are "obviously" waves, but both can be described together—as *laves*. Let us rejoice in paradoxes. Through them, we transcend the "obvious."

14

Laves

There exists a body of exact mathematical laws, but these cannot be interpreted as expressing simple relationships between objects existing in space and time.
—W. Heisenberg, *Physical Principles of the Quantum Theory* (1930)

In 1907, G. I. Taylor, later to be Sir Geoffrey, performed a remarkable experiment. He wanted to see to the heart of our paradox by forcing a direct confrontation between the lumpy and wavy behaviors of light, to catch nature in the act of somehow making lumps interfere like waves.

Turning back to figure 12.1, you will see the setup for the classic sort of wave-detection interference experiment, two slits illumined by a single source. Taylor's innovation was to use such a feeble source that the lumpiness of the light it emitted became apparent.

To make his feeble source, Taylor put the light of a gas flame through several layers of smoked glass and then through a narrow hole. "A simple calculation will shew," he says in his paper, "that the amount of energy falling on the plate . . . was the same as that from a standard candle burning at a distance slightly exceeding a mile." To get a clear picture from this light, he had to expose his film for three months.

Taylor's source was so dim that only one photon at a time was passing through his apparatus. And yet—he found the same interference pattern he would have gotten had he been using a much brighter source.

The more you think about this, the odder it seems. It is the central oddity of quantal physics. If you close one of the two slits, the interference pattern vanishes. We would say, if we imagine waves of light spreading out through space, that the pattern depends on light from one slit having something to interfere <u>with</u>, namely, light from the other slit. But if photons pass through the apparatus one at a time, what can be interfering with each isolated photon?

The inescapable conclusion is that each photon interferes with itself.

This, however, is a most bizarre thing for a lump to do. Consider, for example, what happens at a *node,* a dark spot where (because of destructive interference) no light at all reaches the screen. If we close one slit, removing the interference effect, the dark spots are no longer dark. Through one slit, the node receives some light; through two slits, it gets none. By giving the photons <u>more</u> pathways to the goal, you get a <u>lesser</u> result.

Ordinary particles certainly don't behave this way. Any lump worthy of the name passes through either one slit or the other, and so if both slits are open, the probability of arrival with both slits open is just the sum of the probabilities for each slit separately. Photons, however, do not follow this rule.

Why can't we just <u>look at</u> each photon to see which slit it went through? (A fraction of a photon is never seen, so it can't be going through both.) Surely we can't get interference then, since (for instance) destructive interference requires two canceling contributions, one from each slit. This seems to be an outright contradiction, not just a paradox. But physical events cannot contain contradictions. There must be a mistaken assumption somewhere. To locate it, let's think carefully and in detail about how we would try to observe our contradiction.

First of all, how do you <u>look at</u> a photon? The simplest way, of course, is to use your eye. But wait a minute—you <u>see</u> the photon only when it is absorbed in your retina and induces chemical changes there. So if you <u>see</u> the photon, your eye has absorbed it. If you observe photons this way, there is certainly no contradiction. The act of observing the photon destroys it. There is no interference pattern, because there is no light.

We can try a more flexible version of the same thing, by putting immediately behind each slit little detectors that absorb some but not all of the light going through them. When a photon is absorbed, we learn which slit it passed through. But we quickly realize that this doesn't help. A photon either is detected or isn't—remember, photons come in lumps. If it is detected, it has been absorbed, destroyed, and so it doesn't contribute to the pattern on the wall. If it isn't absorbed, it leaves no energy in either detector and we haven't learned anything about its position. The result? We still find the same interference pattern, just not as intense (because some photons are lost).

If we don't actually observe the photon, we can't say which slit it went through. This much is hardly shocking. The new wrinkle is that we get into contradictions if we even imagine that the photon goes either through one slit or through the other. If we observe it, we find that it does go through one or the other. But if we don't, it doesn't. We must learn to live with the fact that we cannot discuss what "really" happens if we don't specify a procedure for observing it. Which slit does the photon "really" pass through when we don't observe it? Not only don't we know, but even posing the question is an insidious temptation to error. Wittgenstein said it best: "Whereof one cannot speak, thereof one must be silent."

Is there perhaps some more subtle way to do the observation? Let's try the following. Electrons are charged particles, so they interact with photons and can collide with them and be deflected. The photons, of course, are also deflected. Let's try to observe the photons and check which slit they went through, by putting a beam of electrons across the slits and seeing how they get deflected. In this way, we can get some idea of where the photon is, without destroying it.

Now, if the electron gets deflected by a large amount, it has exchanged considerable energy with the photon. The trajectory of the photon is therefore changed. We can learn the position of the photon, but only at the price of disturbing it. When this procedure is followed, no interference is observed. But, of course, we have changed the experiment in an essential way. We have not found a contradiction; we have merely changed the subject. We have interfered with the interference.

More interesting is what happens if the electron gets deflected just a little bit. Offhand it would seem that a little deflection is as good as a big one. By carefully measuring the position and direction of the electron after it has scattered, we should be able to trace its path backward and calculate exactly where the scattering took place. Pocket billiards players must make a closely related calculation anytime they want to bank a ball into the pocket: knowing the desired trajectories, they have to calculate where to do the banking. And this method actually would work, landing us directly in our contradiction, if electrons behaved like billiard balls.

So what gives? The wonderful resolution is that electrons exhibit the same peculiar behaviors that we found for light. They do not behave like ordinary billiard balls, particles, or lumps. We have learned that we cannot say the photon lave passes through one slit or the other, if we do not give a procedure for determining which it is. Once we have had our intuitions shattered in that case, we should be put on the alert. We can no longer take it for granted that we can determine an electron's location, or even its path, without changing in the process of measurement the very things we set out to measure. And if we are to avoid our contradiction, it must in fact be impossible, as a matter of principle, to do the ideal measurement. If we could determine which slit the photon went through—whether by means of electrons, neutrinos, gravity waves, or anything else—without disturbing it, then it would be impossible for the photon to interfere with itself. But Taylor's experiment has taught us that it does.

Our frustration, in trying to measure things without disturbing them, is therefore not merely a passing weakness of technology. It is a fundamental principle of nature, a form of Heisenberg's celebrated uncertainty principle. And analyzing how the frustration arises in simple experiments strongly suggests that not only light but also electrons and, indeed, all other forms of matter behave in similarly peculiar ways. Something as peculiar as we find light to be can interact consistently only with equally peculiar stuff.

Having convinced ourselves that the strange behavior of light at least does not involve logical contradictions—just dancing on the knife-edge of logic—we can begin to discuss it rationally. Light comes in lumps but interferes like waves. Although each metaphor captures some of the reality of light, neither suffices. There is no special word for stuff that behaves the way light does, but one is certainly needed. So we have invented one. We say that light behaves like *laves* or, in other words, that photons are laves.

Laves are mathematical waves that describe the probability for finding particles. To understand this idea, let's consider some examples. Let's see how the lave description works for Taylor's experiment. There is a mathematical wave impinging on both slits. We need something at both slits in order to get interference. But if we observe the photon, we find that it passes completely through one slit or the other, never a little bit of each. So what does our mathematical wave describe? As Heisenberg remarks in the epigraph of this chapter, it does not describe an object (in any ordinary meaning of this word) existing in space and time. The only object around is the photon, and our mathematical wave is certainly not telling us how the photon is spread around. In fact, the photon is *never* observed to be spread around in space—it is a lump. The only consistent interpretation that anyone has ever been able to assign to the mathematical wave is that the mathematical wave describes the *probability* for finding a photon lump. (The precise relationship is that the probability for finding a photon at any point is proportional to the square of the amplitude of the wave at that point.)

For another example, consider the most basic radiation process, a single atom emitting light. In a wave model, the emission is pictured as a disturbance spreading through space, in the form of a sphere centered at the atom expanding outward at the speed of light. In a particle model, a lump of energy goes off in some definite direction. The lave description combines these two. A mathematical wave emerges from the atom and expands at the speed of light. This wave, however, is not itself an ordinary physical disturbance, substance, or object. Rather, it is a table of odds. It specifies the *probability* for finding a lump of light. An observer will always find one photon, moving in some particular direction, but the spherical lave means that any direction is as likely as any other.

Finally, we must mention one other sort of example. We may have probabilities in *time* as well as in space. A clear and extremely important case of this is the phenomenon of radioactive decay. A sample of radioactive material contains millions of identical nuclei. Here we mean identical in a very literal and precise sense, as will become clear in chapter 16. A basic feature of radioactivity is that every second a certain fraction of these nuclei spontaneously break up, emitting various kinds of energy and radiation. Exactly which nuclei will break up is not predictable—remember, all the nuclei are exactly identical, so there can be no basis for such a prediction. The best physics can do, here again, is to calculate a table of odds.

In the description of Taylor's experiment, we had a lave with two branches penetrating the two slits. Each branch describes a photon passing through one of the slits. In the description of a radioactive nucleus, there is a wave branching in time. On one branch the nucleus remains intact; on the other it breaks up. When we observe, we find the nucleus on one branch or the other. But unless the observation is actually made, both branches must be regarded as open, no less for radioactive nuclei than for light. The wave in either case is a mathematical construct that embraces two incompatible alternatives, assigning to each its probability.

It once seemed clear that light was basically continuous and wavelike; matter, basically discontinuous and lumpy. While examining the strange behavior of light, we have seen that to avoid contradictions electrons—or anything else with which light interacts—must also behave in similarly strange ways. We can now begin to appreciate the origin and the meaning of the central doctrine of quantum physics, which is simply that <u>everything is waves</u>. For example, it is possible to do experiments like Taylor's but using <u>electrons</u> in place of photons. And it is found that, sure enough, electrons make interference patterns. So do neutrons.

In fact, X-ray, electron, and neutron interference have all become commonplace techniques for probing the internal structure of molecules. Within a molecule, the individual atoms all act to scatter the incoming wave, so each atom mimics a source of waves, just as does a slit in Taylor's arrangement. Information about the atoms' positions and properties shows up encoded in the resulting interference patterns. For example, Rosalind Franklin's X-ray interference patterns for DNA were critical clues for determining its double-helix structure; but this is only the most dramatic case among many thousands.

The laws of quantum mechanics are not <u>only</u> a description of the microworld, to be replaced with something different for large bodies. They are supposed to describe everything, including large bodies. We may not be too disturbed to learn that an accurate description of electrons and neutrons requires us to regard them as waves, although they appeared at first and are usually referred to as particles. But what about billiard balls? What about cats? And how about you? Do you feel like a wave—uncertain which way to turn, until someone observes you? A necessary part of understanding the strange world of quantum physics, filled with waves and saturated with chance, is understanding how this world could possibly manage at the same time to be our own, not so strange world. To this task, we now turn.

15

Branching Worlds

[W]henever a creature was faced with several possible courses of action, it took them all, thereby creating many distinct histories of the cosmos. Since in every evolutionary sequence of the cosmos there were many creatures and each was constantly faced with many possible courses, and the combinations of all their courses were innumerable, an infinity of distinct universes exfoliated from every moment of every temporal sequence in this cosmos.

—O. Stapledon, *Star-Maker* (1937)

It is notorious that many of the founders of quantum mechanics, including Planck (as in Planck's constant), Einstein (the inventor of the photon), Louis Victor de Broglie (the first to suspect wave properties of electrons), and Erwin Schrödinger (the author of Schrödinger's equation, the basic equation of quantum mechanics), never fully accepted the theory. Their objections did not stem from any problematic experimental results; indeed, since the basic ideas crystallized, in the 1920s, quantum mechanics has gone from triumph to triumph. Some of its triumphs are documented in this book, and others sit inside the Macintosh on which it was typed.

Rather, these very clever and hard-thinking people, each of whom had wrestled with the paradoxes and apparent contradictions of quantal reality, did not feel comfortable dancing on the knife-edge of logic and were deeply troubled by the entry of chance into the basic laws of physics. They felt that something more robust, some <u>real</u> reality, was lurking beneath and that laves, with their fundamental element of chance, were merely a temporary crutch.

Just as when we look we are equally likely to find a photon going through either slit in Taylor's experiment, so when we observe the world developing in time there are a variety of possible outcomes. Laves branch in time, and the branches describe different possible outcomes of observations.

Quantum theory requires that we renounce the possibility to predict the results of individual measurements, even in principle. It tells us that the future <u>cannot</u> be predicted uniquely from the past. Instead, there is a variety of possible futures, each with its own probability. The old giants were not prepared to make this renunciation, to give up on predictability.

After an initial period full of fruitless attempts to devise experiments that would definitely contradict the ideas of quantum mechanics, the objections took a more emotional and philosophical tone. Einstein many times affirmed, "I cannot believe that God plays dice with the world"; in a related context, he said, "This balancing on the dizzying path between genius and madness is awful." And Schrödinger introduced his cat.

Schrödinger's cat is the subject of an imaginary experiment, designed to make orthodox quantum theory seem, not necessarily wrong or contradictory, but ridiculous. The cat is in a sealed box. Also in the box is a little jar of poison gas and a fiendish device to release the poison. The device is triggered by the decay of a single radioactive nucleus. That is, if the nucleus emits an alpha particle, it will set off a firing mechanism that opens the jar of poison. Thus the fate of the cat hinges on a microscopic event, which is known to be governed by the rules of quantum physics. It is a matter of chance, described by a lave that contains both a "live cat" branch and a "dead cat" branch. As time goes on, the "live cat" component gets smaller and the "dead cat" branch gets bigger, because it becomes more likely that the decay has occurred.

Now, the rule of orthodox quantum mechanics is that if we don't make an observation, both branches coexist. But when we open the box, of course, we will find either a live cat or a dead cat (and not a combination of the two). Is it reasonable, Schrödinger asks, that the act of opening the box so drastically alters the nature of the cat? It is one thing to consider how <u>looking at</u> a photon may disturb its behavior. After all, we don't have much direct experience with single photons. But it seems quite ridiculous to think that just <u>looking at</u> a cat could change it much. The idea flies in the face of all our experience with cats.

Eugene Wigner carried Schrödinger's experiments a step further.

He imagined a person in the box instead of a cat. The person is called "Wigner's friend" (although subjecting him to this experiment is no way to treat a friend). Now, suppose that after waiting a while, so there is some reasonable probability that the radioactive decay has occurred, we open the box. In favorable cases, we will find Wigner's friend inside, alive. Then we can ask him, "How did you feel <u>before</u> we opened the box?"

Suppose he answered something like, "Well I might have been dead and I might have been alive; I couldn't be quite sure. But when you opened the box, it suddenly became quite clear to me that I was alive." One possible reaction might be to admire and envy his direct intuitive experience of quantal uncertainty. A much more reasonable reaction, though, would be to doubt his veracity or his sanity. We find it impossible to imagine ourselves being in what Wigner's friend describes as his state of consciousness. We cannot conceive of a mental state that corresponds to the reality of quantal uncertainty.

Schrödinger's cat and Wigner's friend are both imaginary beings devised to embody apparent problems with orthodox quantum mechanics. Schrödinger argued that essentially new physical principles must be invoked to describe macroscopic objects like cats; Wigner concluded that quantum mechanics is inadequate to describe the phenomenon of consciousness.

The orthodox response to these challenges, which emerged from long discussions between Niels Bohr and the younger generation of quantum theorists who passed through Bohr's institute in Copenhagen during the 1930s, is called the Copenhagen interpretation of quantum mechanics. The fundamental move in the Copenhagen interpretation is to make a sharp distinction between a "quantal" realm and a "classical" realm. Microscopic objects like photons or electrons clearly belong to the quantal realm. Laves—bringing with them uncertainty and an element of chance—are needed to describe their behavior. On the other hand, macroscopic objects, such as the dials and pointers physicists typically use to read out the results of their experiments, or the physicists' pet cats and friends, belong to the classical realm. They can be described by classical (that is, prequantal) physics. We do not need laves, uncertainty, or chance to describe them. In fact, it seems we <u>cannot</u> use these ideas without running into the sorts of paradoxes that Schrödinger and Wigner raise.

This division of the world into two distinct realms, however, im- mediately forces new questions. First of all, precisely where is the line drawn between them? This has never been made completely clear. In many examples, it is obvious which realm you are in. For instance, a single photon, described by a lave, is in the quantal realm. But for us to sense the photon as light, the photon must trigger complicated chemical and electrical changes in our heads, and somewhere along the line it enters our consciousness. Our con- sciousness must be in the classical realm, as Wigner's friend has taught us. But when exactly did the change occur? And what is the nature of the event that made it possible for the different realms, governed by different laws of physics and even different logics (chance versus determinism), to make contact?

In authoritative statements of the Copenhagen interpretation, one finds sentences like the following: "Experience only makes state- ments of this type: an observer has made a certain (subjective) obser- vation; and never any like this: a physical quantity has a certain value" and "The laws of nature which we formulate mathematically in quantum theory deal no longer with the particles themselves but with our knowledge of the elementary particles." In light of such statements, it seems fair to summarize the basis of the Copenhagen interpretation in terms of two main principles:

1. In science, we can and should deal only with statements that have some direct connection with what we can experience.

2. We cannot directly experience quantal reality (as our reaction to Wigner's friend taught us). It "exists" only as a mathematical construct and takes on definite meaning through its influence on a separate classical realm. The exact boundaries of the classical realm are left unclear, but this realm clearly must include everything in our conscious awareness.

The interface between classical and quantal realms is supposed to take place by a process called *collapse of the wave function*. This process is simply explained in the context of Schrödinger's cat exper- iment. In describing the radioactive nucleus at the heart of this experiment according to quantum theory, we must use a lave that has both "nucleus intact" and "nucleus disintegrated" branches. In our previous discussion of this experiment, we said that these two branches would lead to "cat alive" and "cat dead" branches for the lave describing the cat. However, the Copenhagen interpretation would describe things differently, as follows.

The cat, being a macroscopic object, belongs in the classical realm. It is not described by a lave. It is either alive or dead, for sure, although if we don't make an observation, we won't know which alternative holds. At some stage between the quantal nucleus and the classical cat, a special process took place. In the quantal realm, there was a lave containing two branches; when this realm interfaces with the classical realm, only one branch is selected. (We might think of the different branches of a lave as so many sperm, which all compete to be first to reach the egg of classical reality. Only one does so, and only this one gets to realize its potential and influence the subsequent larger events.) Which branch of the lave gets selected is determined by chance, but only one does, and only this one influences the unique subsequent evolution of the classical realm.

The step whereby all possibilities but one are annihilated, as the branches of the lave are pruned down to a single one, is the collapse of the wave function we mentioned earlier. It is absolutely necessary to have such a step if the quantal realm, aboil with uncertainty and chance, is to make contact with a far more ordered and determinate classical realm.

In this way, the Copenhagen interpretation avoids the difficulties with Schrödinger's cat and Wigner's friend. Quantal behavior for cats and people, which is never observed and seems ridiculous, has been removed. However, this victory has been bought at a very high price. The device of introducing two realms of reality seems extremely artificial, and the process of collapse of the wave function, whereby these two realms make contact, is both vague and arbitrary. Although the Copenhagen interpretation presents an ingenious and logically possible way of reconciling the success of quantum theory in the microworld with its apparent absurdity when applied to large objects, many physicists remain dissatisfied with it.

For those who remain dissatisfied with the orthodox Copenhagen interpretation, there are several alternatives. One possibility is to conclude that quantum mechanics is fundamentally flawed. Perhaps fundamentally new effects operate when very many particles are involved. These effects would have to be tiny for nuclei and atoms, so as not to disturb the many quantitative successes of standard quantum mechanics. It might be possible that they add up for large bodies and work to make these bodies behave more "classically." If these ideas could be made to work, it might be possible to pass smoothly from the quantal realm to the classical realm, without having to make a strict division between these realms or to invoke the mysterious process of collapse of the wave function to let the two realms communicate. A few physicists espouse this approach and are actually trying to make theories of how quantum mechanics might be modified and even devising experiments to test for deviations from it. Thus far, however, nothing tangible has emerged from their efforts.

Another possibility, which we think is much more attractive and which leads to exciting consequences, is to adopt a <u>radically conservative</u> attitude. That is, we follow out quantum mechanics utterly seriously, as it stands and without any alteration, to see where it leads.

Right away, we find ourselves swept away from comfortable Copenhagen. Our new point of view comes into conflict with the orthodox Copenhagen interpretation, on several fronts. The basic equations of quantum mechanics are all written in terms of laves. There are not two separate realms in these equations; the equations simply do not contain a separate classical realm. Nor do the equations contain anything that corresponds to a collapse of the wave function. The equations for laves are perfectly definite, determinate equations. They do not contain any special terms for cats, consciousness, or observers. And branches never get pruned away.

It would seem, then, that we have put ourselves in grave danger of paradox or outright contradiction. If the quantal world of laves <u>is</u> our universe, the contradictory possibilities contained in the laves must all be counted as parts of reality. In addition, we are left exposed to the challenge posed by the paradoxical quantal descriptions of Schrödinger's cat and Wigner's friend.

Nevertheless, it does seem possible to maintain a radically conservative stance, and to take our quantum mechanics straight. Let's see how.

To get to the root of the problems of interpreting quantum theory, we must step back for a moment to discuss the way an inherently probabilistic theory like quantum mechanics is used. Again, to keep things as concrete and simple as possible, let's talk specifically about Schrödinger's cat experiment. Quantum mechanics allows us to calculate how likely it is that when we open the box, we will find a dead cat. But what is the value of such a lame "prediction"? The cat will be either alive or dead, period. We couldn't tell beforehand which it would be, and doing a calculation in quantum mechanics doesn't change this situation. So how has the theory helped us?

The theory is certainly useful when the odds are overwhelming. If the cat is exposed for only a short time, so short that the probability that the nucleus will decay is say less than one in a billion, then we would have a right to be very surprised if we looked in and found a dead cat. After a few surprises like this, we would give up on the theory. Similarly, if the cat is exposed for a long time, so long that a calculation in quantum mechanics tells us there is virtually no chance the nucleus has escaped decay, then, if we believe in the theory, we must be confident that when we look in we will find a dead cat.

So if the odds are overwhelming, it is clear how the theory is useful. But how can we make use of the theory in the usual case, when the odds are <u>not</u> overwhelming? The easiest and most straightforward way is to look at a <u>large number of identical cases</u>. Suppose we fix the exposure time so that, after being exposed, the cat is equally likely to be alive or dead. If we study just one cat, we cannot make very good use of this information. But if we expose ten thousand cats, then the theory tells us that some things become overwhelmingly unlikely. For instance, we would become very suspicious of the theory (or the experiment) if all ten thousand cats were found to be alive or if all ten thousand cats were found to be dead. In fact, the odds that more than 5,600 or fewer than 4,400 cats are alive is less than one in a billion, so even this much of a deviation would raise our eyebrows very high. The point is that we can use close odds for a single system to generate overwhelming odds for a large number of similar systems. And there is no subtlety about how to interpret overwhelming odds—we treat them as certainties.

It is neither practical nor humane to carry out this cat massacre (not to mention the extension to Wigner's friends). The principle involved is quite general, though. For instance, we can discuss Taylor's experiment in the same terms. For each photon, quantum theory predicts a probability distribution. We cannot tell where any particular photon will arrive on the viewing screen (that is, the photographic plate placed behind the slits). However, a great many photons are involved in producing an image we can see on a photographic plate. When we add up the probabilities for so many photons, we get a pretty definite average result, and significant deviations from this average are extremely unlikely. From what might seem a loose and wishy-washy description of individual photons, we generate quite a precise prediction for what the interference pattern on the photographic plate—the total result from many identical photons—looks like.

Nature provides us with many other examples where chancy predictions of quantum theory for individual atomic or nuclear processes generate overwhelming odds, or effective certainty, for observable processes. The fundamental reason why this situation is so common is that there are enormous numbers of atoms and nuclei in what we ordinarily think of as small bits of matter. We have rhapsodized on this theme before. And here we once more find reason to admire the immensity of $\mathcal{N}$; it allows us to pass smoothly from quantal chance to effective certainty.

The fact that large numbers of probabilistic elements may behave in a determinate way begins to show us a path toward making some contact with the ideas of the Copenhagen interpretation. That is, what is called the "classical realm" in the Copenhagen interpretation is already implicit in pure quantal reality. Macroscopic objects containing very large numbers of laves from the "quantal realm" will behave in very nearly determinate ways, following the averages, just as a matter of statistics.

We can accordingly give a sensible account of macroscopic objects and their interactions with each other, within a pure quantal reality. The typical quantal features of uncertainty and chance, which are so contrary to our intuitions for ordinary objects, need never appear. More subtle is the question of how small objects, which do exhibit uncertainty and chance, interact with large ones that don't. In other words, we must face up to Schrödinger's cat experiment.

We have taken the first step, by understanding how cats that behave in something like the way we expect of cats—that is, that they move in definite paths in space, they do not fluctuate between life and death, and so forth—could exist in a purely quantal world. But what of the strange cat lave that inhabits the box in Schrödinger's setup, that half-and-half mixture of life and death? The complete lave of the universe must contain both a "live cat" and a "dead cat" branch. It must also contain, for the observer who looks at the cat, both an "observed live cat" and an "observed dead cat" branch. The simple but crucial point is that these branches are *correlated*. One branch of the lave of the universe contains both the "live cat" description of the cat and the "observed live cat" description of the observer, while another branch contains both the "dead cat" description of the cat and the "observed dead cat" description of the observer. The branches separate when a radioactive decay occurs. The moment when radioactive decay occurs is a matter of chance, but once it occurs it triggers reactions in larger bodies (the bottle of poison, the cat, the observer). Large numbers of particles become involved, and the later steps no longer contain significant elements of chance. Once the bottle of poison is broken open, the cat will surely die, and once the cat dies, we will surely see a dead cat if we look in. The evolution of the macroworld is in no way strange, chancy, or paradoxical. Yet, this evolution can be put in motion by strange and chancy, but we hope no longer paradoxical, quantal events.

Or, to put it more dramatically, different worlds—including diverging histories for what we think of as ordinary or "classical" macroscopic objects—branch out from the initial quantal lave. *The total lave of the universe contains descriptions of many possible worlds, all on an equal footing. When we observe Schrödinger's cat, we do not change its state. Rather, we find out which world we belong to.*

From a radically conservative approach to quantum mechanics, we reach astonishing conclusions. We find that there is no experimental or logical need to alter the framework of the theory. In particular, there is no experimental or logical barrier to having chance at the foundations of physical reality, despite appearances. Nor are external concepts, such as the separate "classical realm" of the Copenhagen interpretation, required. Large objects affect one another in thoroughly predictable ways—that is, they behave as if they belonged to a "classical realm"—despite the underlying quantal microworld, for simple statistical reasons. Life insurance companies can rely on their return, because individual differences in life span are submerged in the average over many individuals. Ordinary physical objects contain far more atoms than the human population of earth, and you can bank on their following the average.

Most important, by taking seriously the idea that the universe is described by a lave, we are inevitably led to consider that it contains many alternative realities. This is the price of doing away with the peculiar notion, alien to the orderly equations of quantum theory, that wave functions "collapse." If the lave, or wave function, of the universe does not "collapse" when something from the "classical realm" observes it—and we think it is beautiful and wise to do without both these notions—then this lave must continue to contain every possibility. Stapledon thought he was describing a fantastic imaginary cosmos, but he was in fact describing the quantal reality of our own universe. Our universe is described by a lave with many branches, many distinct worlds, each giving birth to many fertile daughters as time unfolds.

Why isn't the many-branched nature of quantal reality, the reality underlying our universe, more evident to us? Why do we have to infer it by subtly interpreting observations on microscopic objects? Why can't we feel the presence of the other worlds? It must be because our consciousness is based on processes involving very many atoms. It therefore is sensitive only to coarse average behavior of the microworld. Or, to put it another way, our brain functions as if it belonged in the "classical realm," and it effectively collapses incoming wave functions. Although there is at present no genuine understanding of the physical basis of consciousness, it seems quite plausible that it is based on the concerted action of many atoms indeed. Our brains evolved to respond to cues from macroscopic objects— such as potential food, predators, and other human beings—in a reliable way. Quantal uncertainty is less than useless in these tasks. So it is very reasonable a priori, as well as obviously true from experience, that we are not directly aware of quantal reality. If this interpretation is right, it means that the physical processes that underlie consciousness must involve correlations between aggregates of many molecules, correlations that are only slightly, if at all, affected by quantal uncertainty.

People generally live out their conscious lives on one branch of the lave of the universe, unaware of the other branches. But once we accept a radically conservative interpretation of quantum mechanics, we are haunted by the awareness that infinitely many slightly variant copies of ourselves are living out their parallel lives and that at every moment more duplicates spring into existence and take up our many alternative futures.

In what ways, concretely, do the branching worlds differ? To put this question another way; What specific important events are in-

fluenced by quantal uncertainty? It is macabre fun to consider imaginary experiments with Schrödinger cats, but does anything
remotely similar ordinarily happen in the natural world? The answer is—yes, certainly. An important example is biological mutations. Such mutations can be caused by the impact of a single ultraviolet photon on a DNA molecule, which disrupts the molecule's
structure. Now, whether a particular photon interacts with a particular molecule is a matter of quantal chance. So here is a specific
instance where the quantal element of chance can have a critical
influence on larger matters.

It is said that the history of the world would be entirely different
if Helen of Troy had had a wart at the tip of her nose. Well, warts
can arise from mutations in single cells, often triggered by exposure
to the ultraviolet rays of the sun. Conclusion: there are many, many
worlds in which Helen of Troy <u>did</u> have a wart at the tip of her nose.

Frustration and Uncertainty

Scientific laws come from many sources and take many forms. It is amusing that some of the most important laws arise from *frustration*. When scientists work very hard to accomplish something but have no success, they start to suspect that nature is conspiring against them.

Here are three examples of this process:

They try to build a perpetual motion machine—and become frustrated. This very frustration becomes the second law of thermodynamics: it is impossible to extract work from an isolated system.

They try to catch up with a light beam—and become frustrated. This very frustration becomes the principle of special relativity: it is impossible to change the speed of light by your motion.

They try to pin down simultaneously the position and velocity of a particle—and become frustrated (the contribution of quantum mechanics to this genre). This very frustration becomes Heisenberg's uncertainty principle: it is impossible to measure simultaneously the position and the momentum of a particle accurately. The product of the uncertainty in position and the uncertainty in momentum will always exceed Planck's constant.

Why do physicists seem to enjoy formulating physical laws in terms of their frustrations, as statements of what they can't accomplish? Is this a useful form or just a consolation to wounded pride? ("If we can't do it, it can't be done.") No doubt, part of the charm of formulating laws of physics as frustration principles is the consolation it gives to frustrated egos, but there are more hardheaded advantages too.

For one thing, frustration principles give practical advice: "Don't waste your time on the impossible."

If we accept this advice, we liberate ourselves from practical frustrations and turn with a good conscience to the enchanting realm of *imaginary,* or "thought," experiments. For example, once we admit the futility of chasing light beams, we turn (in relativity theory) to filling space with imaginary rocket ships and working out the consequences of the fact that all of them measure the same speed of light. Similarly, once we admit the futility of building real machines with perpetual motion, we turn (in thermodynamics) to imagining Rube Goldberg contraptions of wild ingenuity and to working out the consequences of the fact that they can't produce perpetual motion. When action is systematically frustrated, thought delights in understanding precisely how action is frustrated, case by case.

Finally, there will always be a few stubborn souls who refuse to believe the frustration principles. These few reject the practical advice ("Heavier-than-air machines will never fly; don't waste your time trying") such principles offer. Usually, their efforts do turn out to be a waste of time. But at least the principles direct their aim at important targets—and remember the *usually* in Treiman's theorem.

We ran head-on into a frustration principle in our discussion of Taylor's experiment. If we could tell which hole a photon went through, or even determine that it went through one or the other, the result of that experiment—the interference pattern—could not be what it is. So there is a frustration principle lurking here: efforts to look too closely into what the photons are doing are doomed to frustration.

By considering precisely what goes awry in various plausible efforts to relieve this frustration, Werner Heisenberg arrived at a general diagnosis: that it is impossible simultaneously to make accurate measurements of both a particle's position and its velocity. The more accurate you make your measurement of position, the more you disrupt the particle's velocity—and vice versa. This statement (in numerical form) is the celebrated *Heisenberg uncertainty principle.*

The Heisenberg uncertainty principle is a prime example of the positive benefits arising from frustration. For one thing, it puts a solid limit on just how much you can make a lave look like a particle.

This limit is what saves the Taylor experiment from rampant self-contradiction. If we <u>could</u> measure perfectly both the position and the direction of the photons as they moved through the holes, we would be able to predict the distribution of light on the screen—and the predicted distribution would be wrong, with no interference. Heisenberg's principle saves us. If we measure the position accurately, we change the direction of motion; so then it is not surprising that the interference pattern is disrupted. In a similar way, we find that our Taylor experiment frustration leads to uncertainty for electrons or for anything else we can bounce off photons—that is, just plain everything.

The uncertainty principle also frustrates electrons in atoms, to our benefit. Consider the hydrogen atom—an electron and proton bound together by electrical attraction. Since the electron is attracted to the proton, its inclination is to snuggle up as close as possible. Why doesn't the electron just settle down right on top of the proton? If that happened, atoms would be as small as their nuclei, and chemistry as we know it would not exist.

Once again, Heisenberg's uncertainty principle comes to the rescue. If the electron were localized in a small region of space, such as on top of a nucleus, then according to Heisenberg it would have a large spread in velocities. Large velocities mean large energy. Thus, for an electron, the attractiveness of snuggling up to a nucleus must be balanced against the enormous amount of energy required to "settle down" in such a constricted region of space. Electrons remain somewhat frustrated, and atoms don't collapse.

The uncertainty principle is born from our frustrations, but it grows into a versatile tool of positive scientific understanding. By blaming our frustrations on nature, we end up making solid progress in understanding her. Mild cosmic paranoia seems to be a fruitful attitude.

A Quantum Lottery

People react in different ways to the idea that the universe contains a diverging infinity of branching worlds. Many recoil from this infinity. They feel that all their efforts, and the individuals and things they cherish, are belittled by the existence of many alternative realities. Others are aesthetically appalled at the extravagance. "Why are there so many worlds," they ask, "when one would do?" Some are annoyed that they can be conscious of only an infinitesimal slice of reality. They feel cheated of all the fun and interesting things they are missing.

We have passed through each of these moods. On the whole, though, we have come to enjoy the idea of branching worlds, and we heartily recommend them. Have you made mistakes in your life? Do you have fantasies about different roads you might have taken? Did you pick the wrong college, marry the wrong person, or choose the wrong profession? Or do you, perhaps, have a wart you don't like? It is a comfort to think that there are worlds in which each of these things is different.

We have even discovered a promising application of quantal reality to increasing the sum of human happiness—a sure path to wealth for everyone, courtesy of branching worlds. In fact, we can guarantee that everyone gets to become a millionaire by participating in a quantum lottery—in which everybody wins!

It is a simple matter to design such a lottery. For instance, we could sell chances on when some particular radioactive nucleus will decay. Let a million people place one-dollar bets, and let the person who has the right time collect the full sum. Now, according to our interpretation of quantum mechanics, every possible decay time is realized, in some world or other—and so everybody gets to be a millionaire (in some world or other).

Sixth Theme

Radical Uniformity in Microcosm

Having discussed quantal reality, we are ready to approach the uniformity of the microworld in a fundamental way.

In the microworld, we meet uniformity of the strongest kind: complete indistinguishability. A remarkable, direct, and absolutely precise way of testing whether two bodies—say, two atoms—are truly indistinguishable will be described. Atoms are often found to be utterly indistinguishable from one another. This fact presents riddles at two levels.

Even given that the atoms are made from identical building blocks, how does nature assure perfect "quality control" in assembling them? We shall find a beautiful answer, closely related to the physics of musical instruments.

And then, <u>why</u> are the building blocks identical? It will emerge that two indistinguishable particles—for instance, two electrons—are identical because both are manifestations of a single underlying reality, a world-filling *field*.

Prelude Six

BACK TO PYTHAGORAS

Human history has witnessed the fall of many religions, but perhaps none fell for stranger reasons than did the Pythagorean. Its fall had to do with the length of the diagonal across a square.

The Pythagoreans were a community of mathematical mystics who flourished in southern Italy during the fifth century B.C. Their doctrines had an enormous influence on later Greek philosophers and, through their writings, on the founders of modern science. A keystone of their world system was the conviction that "all things are numbers." This seminal idea originated with Pythagoras himself. It arose in part from his marvelous discovery that the musical tone produced by a plucked lyre string is related in a surprisingly meaningful way to its length. Anyone who plays with a stringed instrument for more than a few seconds will notice that the shorter the string, the higher the pitch of the note it produces. Pythagoras went a small, but epoch-making, step beyond this observation. He quantified the sensation of harmony. He found that this psychological effect could be measured in numbers. For example, if one string is twice as long as another, the tone it produces is precisely an octave lower. If the lengths are in the ratio of three to two (that is, if one string is three-halves the length of the other), the tones they produce make the musical interval of a fifth. More generally, it is just when the lengths are in ratios of small whole numbers that the tones they create "harmonize" in a pleasing way.

Pythagoras was much impressed by his discovery, and rightly so. He had found a wholly unexpected, and to this day poorly understood, connection between number and sensation. His discovery inspired, or at least reinforced, his mystical faith that beneath the appearances of the world was a deeper level of reality described by numbers, a faith captured in the credo that he handed down to his disciples: "All things are numbers."

The Pythagoreans, inspired by this vision, made pioneering investigations into mathematics. Among other things, they discovered the famous "Pythagorean theorem" of geometry, which relates the lengths of the sides of right triangles. (The square of the hypotenuse is equal to the sum of the squares of the two other sides.) This very theorem, however, brought on the crisis that undermined their faith, for on applying this theorem to calculate the length of the diagonal of a square they reached a shocking conclusion. They found that the lengths of the side of a square and of its diagonal are incommensurate—that is, the ratio of these two lengths cannot be expressed as a ratio of whole numbers. (It is therefore what we now call an irrational number—an eminently Pythagorean usage.) This means it can't be true that everything is made of numbers; if the side of a square is made of a whole number of units, the diagonal cannot be.

For many years, this fact was regarded as a scandal, a dirty secret to be hidden from the outside world. There is a legend that the member of the brotherhood who first revealed the secret was drowned at sea; whether his fate was sealed by human vengeance or by divine retribution is left unclear. In any case, the pretense could not be maintained indefinitely, and Pythagoreanism in its original form declined.

The vision of a universe governed by mathematical structures of some more general kind, not restricted to whole numbers, was more robust. It was developed by the later Greeks and adopted by the founders of modern science, who regarded themselves as the inheritors of this tradition. Here is Galileo (in his *Starry Messenger*):

Philosophy is written in that vast book which stands forever open before our eyes, I mean the universe; but it cannot be read until we have learnt the language and become familiar with the characters in which it is written. It is written in mathematical language, and the letters are triangles, circles and other geometrical figures, without which means it is humanly impossible to comprehend a single word.

In this more general sense, the spirit of Pythagoras has been incorporated into modern physics from its beginning. However, for a long time it seemed that his original emphasis on whole numbers was completely misguided. The classical physics of Galileo, Newton, and their followers into the twentieth century is based not on whole numbers but on quantities capable of continuous variation.

Most basic among the quantities assumed to be infinitely divisible are space and time. To choose a modern example, if a car's velocity is steady at sixty miles per hour, you can feel confident that it travels a mile in every minute, one-sixtieth of a mile in every second, and so on down the line. A car's speedometer would be pretty worthless if, before it could give you a figure for "miles per hour," it had to count up miles for an entire hour. Fortunately, the velocity—the ratio of distance traveled to time elapsed—can be computed even by comparing a microscopic distance to a minuscule time.

This kind of computation becomes especially important when the velocity is changing. If you want to know your velocity at 3:00 P.M. precisely, for example, you want the instantaneous value of the distance-to-time ratio—the ratio when the time elapsed is as close as possible to zero.

The concept of acceleration—the instantaneous rate of change in velocity—is a pillar of classical physics. It appears in Newton's fundamental law of motion, which states that force equals mass times acceleration. Once again the ratio in question—the change in velocity divided by the time elapsed—is supposed to be computed as the time elapsed shrinks to zero. It is therefore implicit in the very formulation of Newton's law, as in many of the other laws of pre-quantum physics, that time and space are infinitely divisible. The whole numbers, the element of discreteness that so entranced Pythagoras, seemed to have disappeared from the fundamental description of nature.

In recent times, there has been conspicuous movement, along many fronts, back toward Pythagoras. Scientists have discovered several fundamental aspects of reality, including our raw sensory perception of the world, where the superficial appearance of continuity masks a deeper level of discreteness.

Our sensory perception of the world begins with receptors that translate all messages from the outside world—sights, sounds, or smells—into electrical impulses, the common coinage of our nervous system. These impulses are either all or none; each is identical to all the others, and only their pattern conveys meaning. Our brain reads not the world itself but this encoded version of it, a sort of Morse code translation into series of dots and pauses. From this lumpy input, our consciousness somehow manufactures the illusion of its own continuity.

Likewise, beneath the seemingly infinite and continuous variability of plants and animals is a deeper discreteness. The genetic message that carries all the information necessary to construct the organism is stored in DNA molecules. The possible structures of such molecules are very limited. Each possible structure is determined completely by the sequence of nucleic acid residues that occurs along the backbone of the molecule; at each place along the backbone, one of only four possibilities occurs (adenine, cytosine, guanine, thymine—or A,C,G,T). The genetic message is therefore written in a discrete form; it is a finite, though often very long, word in a language of four letters.

Discrete structures have great advantages over continuous ones in storing and transmitting information. We can illustrate two of the most important by considering the example of money, the medium used to store and transmit information about wealth. One great advantage of having discrete units of money—dollars and cents—is that it makes it possible to specify quantities with no ambiguity. Exact quantities of money can be fixed in definite whole numbers of dollars and cents, and two people can easily agree when two such quantities are equal. That's why making change or similar simple transactions needn't involve haggling. It would be very different if gold and silver were used directly as the means of exchange; every transaction would involve weighing operations that would be inherently subject to small uncertainties as well as to subtle tampering with the scales and so forth. Related to this is a second great advantage of a discrete system: its immunity from small errors. A penny remains a penny though it loses its shine or becomes battered, or even if a little material wears off. You either have a penny or you don't. It would be very different if some continuous quantity, such as the shine of a penny, its shape, or its weight, determined its value. Then these degradations would be constantly eroding and confusing its value. The general advantages of discreteness for precise recognition and immunity from small errors help explain why all modern societies have discrete (digital) monetary systems.

For similar reasons, telegraphers send Morse code messages

rather than a closer imitation of human speech, television and newspaper pictures are in reality arrays of dots, and digital computers have come to dominate their analog cousins. These reasons also explain why biological evolution has developed the stereotyped nerve impulse for transmitting information within an organism and a discrete genetic code for transmitting information across generations.

The laws of physics governing the subatomic world also contain a fundamental element of discreteness, and "good use" is made of this "design principle" in constructing the world as we know it. Perhaps the most direct, and certainly the historically crucial, revelation of discreteness in the microworld is the existence of line spectra. Just to remind you, atoms of each chemical element are resonant with light of only a few specific colors (or, equivalently, wavelengths). These colors make up the element's line spectrum—the lines of color that appear when light from these atoms is sent through a prism.

The line spectrum is identical for different samples of a given chemical element, but it differs markedly between different elements. There are no intermediate forms of spectra interpolating among those of the different elements; the possible patterns do not slide smoothly into one another. And so there is discreteness at two levels: in the lines of light picked out from a continuous rainbow of possibilities by each element, and in the discretely different possible overall patterns. It would surely have delighted Pythagoras to learn that in modern physics the spectral patterns are traced directly to the sequence of whole numbers—to the so-called atomic numbers, which count the number of protons in an atom's nucleus.

Discreteness in the microworld carries the same advantages of precision and immunity to small errors that we have claimed for it in engineering and biology. It is the ability of atoms to recognize one another precisely and unambiguously that ultimately underlies the specificity of chemical and biochemical reactions; and it is the discreteness of their internal structure that allows them to assume stable forms in an ever-agitated world.

The design of monetary systems or computers is guided by intelligent agents; the design of nervous systems and genetic messages has been fixed by evolution, which has a sort of intelligence too. It is not so surprising that their designers have chosen discreteness, given its advantages. But what of the microworld? Who or what designed it?

The element of discreteness enters the microworld as a consequence of certain fundamental laws, the laws of quantum mechanics. We cannot analyze these laws any further. In particular, we need not and cannot infer the presence of a guiding intelligence. The laws tell us how the world works but are mute when asked why. We cannot analyze deeper—but we can come to understand the laws, play around with them by doing imaginary experiments, make analogies, make metaphors, and in such ways come to appreciate their strange beauty. Reconciling the continuous nature of space and time inherited from classical physics with the discreteness found in the microworld was a major achievement of our century, summarized in the science of quantum mechanics. In a truly remarkable way, though, some key ideas are already implicit in Pythagoras's work on musical instruments. After all, a musical instrument generates, from continuous vibrations of strings or air columns, a discrete set of tones. The way in which the continuous vibrations of electrons in an atom generate a discrete set of color "tones"—spectral lines—turns out to be, once properly understood, amazingly similar. So, here too, we come, after two and a half millennia, back to Pythagoras.

16

The Indistinguishable

This world might *be a world in which all things differed, and in which what properties there were were ultimate and had no farther predicates. In such a world there would be as many kinds as there were separate things. . . . But our world is no such a world. It is a very peculiar world, and plays right into logic's hands.*
—W. James, *The Principles of Psychology* (1890)

We once visited a very peculiar artificial world—a display of LEGO® toy bricks. LEGO windmills rotated above a garden filled with LEGO tulips; LEGO airplanes emerged from LEGO cloud banks; LEGO families sat in autos spinning their LEGO wheels on a long Sunday drive to nowhere. This miniature universe of complexity and detail had been built up, according to simple rules, from identical copies of just a few kinds of building blocks.

Modern physics seeks to assemble the universe from building blocks like protons, electrons, and photons. Here, as in the LEGO world, the blocks must be identical if they are to be interchangeable and always "fit." The attractive simplicity of this model should not disguise the deep puzzles it poses. How can we be sure that our basic units are truly identical copies? And if they are, why?

If you start from the intuitive sense of *identical*—meaning "having no differences"—you'll never arrive at definitive answers. Suppose you examine two electrons in the most exhaustive way, and find that they are precisely alike in mass, charge, spin, and so on, to the utmost accuracy of your measuring instruments. How can you be sure that tomorrow some new development in technology won't yield more accurate measurements and turn up some tiny difference? You can't prove identity in this negative way, any more than you can "prove" to a four-year-old that there are never any monsters hiding in the closet.

Given the difficulty of proving a negative proposition, what we need is a positive definition of identity. Remarkably, nature herself provides one, embedded in the laws of quantum physics. In chapter 14, we found that when there were two possible ways ("through slit A" or "through slit B") for the same identical event ("photon arrives at center of screen") to occur, there was an interference pattern. But as soon as we tampered with the indistinguishability of the two events—for example, by finding out which slit the photon went through—the interference pattern vanished.

Such behavior can be used to give a sharp new meaning to identity. If two particles are exactly identical—indistinguishable—we can use them to set up identical events, and look for interference patterns. Identical particles therefore interact in a way markedly different from the way they would interact if there were even a tiny distinction between them. It is convenient to speak, somewhat loosely, of special *identity forces* that exist only between indistinguishable objects.

A simple experiment will show the positive effects of identity. Two beams of particles—one heading east, the other west—are fired directly at one another. Detectors are set on either side of the collision site—that is, along the north-south axis. These detectors count the number of particles that scatter into them.

Figure 16.1 gives a rather schematic (and much easier to draw) interpretation of the setup, with six-guns and targets in place of the particle beams and particle detectors. This picture shows only those bullets that bounce at precisely the right angle to hit the targets. In reality, many bullets bounce off at odd angles to miss the targets completely, and most just go blasting right through, without ever hitting another bullet at all. It's the ones that bounce off at right angles that interest us, however, so we'll just concentrate on them.

Suppose we start with beams of two different, but similar, kinds of "bullets," for example, atoms of helium ^{3}He and ^{4}He. The nuclei inside are (2P, 1N) for ^{3}He, and (2P, 2N) for ^{4}He; in each case, the nucleus is surrounded by two electrons. These two types of atoms, two isotopes of helium, present the same face to the world. The only difference between them is the mass of the nucleus, and the nucleus is shielded deep inside a cloud of electrons.

(The near identity of isotopes makes isolating one from another a

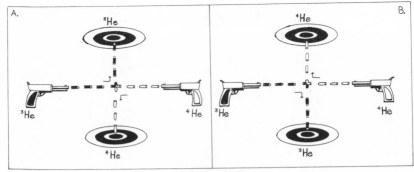

Figure 16.2

laves. Unlike colliding billiard balls, laves cannot be precisely localized; they have a certain spread in position. During the process of
collision, each gets inside the spread of the other, and (for identical
laves) we lose track of which is which. Two identical laves approach
the collision; two identical laves depart. But which is which? We
cannot tell. Each detector records that one lave has arrived there.
Did the lave in the northern detector start out in the east beam or
in the (identical) west beam? For billiard balls, or distinguishable
atoms, we can choose between the two distinct possibilities shown in
figure 16.2. For identical laves, these two events look just alike. It is
here—between indistinguishable events—that the interference
effect arises, just as it did in the Taylor two-slit experiment described
in chapter 14.

Consider two incoming ^{4}He laves, whose paths through time and
space are cloudy and overlapping. Our previous experience with
quantum mechanics has taught us that when there are two alternatives leading to the same result, the probabilities do not simply
add—they interfere. If we could keep track of where each particle
was at every stage, we could tell which route was taken, and the total
probability would be just the sum of the probabilities for the two
routes. But we can't keep track; to measure the position of an atom,
we must disturb it, and this would change the conditions of the
experiment. And if we don't measure, we can't claim to know its
trajectory at all. "Whereof one cannot speak, thereof one must be
silent."

For ^{4}He atoms, we saw that the probability of finding an atom in our
north-south detectors is doubled. Identical events have interacted to
produce the additive wavelike behavior known as constructive interference. Particles that, like ^{4}He, show constructive interference are

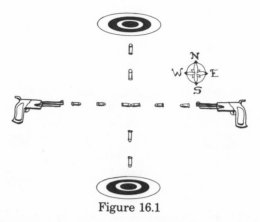

Figure 16.1

very complicated and expensive procedure. In fact, the most difficult step in making an atom bomb is separating uranium isotopes ^{235}U and ^{238}U. ^{235}U is the isotope that's useful for bombs, because it easily captures neutrons and then swiftly breaks up, releasing both energy and more neutrons. On the other hand, ^{235}U always occurs in ores much diluted by ^{238}U, which is worse than useless for bombs. To separate the two kinds of uranium is extremely difficult, because their chemical properties are almost identical. Nature doesn't do it—that's why even in rocks with very different chemical histories the same, small ratio of ^{235}U to ^{238}U is found—and neither can we, at present, short of building costly and elaborate diffusion plants or nuclear reactors. We must be grateful that it is so difficult; otherwise nuclear terrorism would be an ever-present danger. Some of the most dangerous and irresponsible research being done today is aimed at using lasers to make isotope separation much easier.)

Getting back to our beams of helium atoms: when just the right kind of collision occurs to send two atoms bouncing away from each other along the north-south axis and into the two detectors, there are two equally probable outcomes. For a ^{3}He–^{4}He collision, these two symmetric possibilities are shown in figure 16.2. Bringing one beam into collision with the other, we find that each detector picks up scattered atoms at a certain rate—say, 100 atoms per minute of each isotope, for a total of 200 helium atoms per minute.

Next, let's try the same experiment using two beams of ^{4}He. The important ordinary (electric) forces are the same as in the ^{3}He–^{4}He collision, so a first guess would be that each detector still receives 200 helium atoms per minute, 100 from each of the two beams. Instead, the detectors on the north-south axis receive a total of <u>400 atoms</u> of ^{4}He per minute. What's going on?

Remember that, properly speaking, atoms are not particles but said to be *bosons*—a shorthand term for "particles obeying Bose-Einstein statistics." But how do we recognize a boson, short of setting up this special experiment? In other words, what effect—if any—do the peculiarities of indistinguishable particles have in the natural world?

One way to recognize bosons is by their tendency to imitate each other. For instance, in our beam experiment the probability for scattering at right angles is enhanced because of the symmetry of the arrangement—both bosons are getting scattered through 90°; they're doing the same thing. That's what bosons like to do. Similarly, the presence of one boson increases the chance that another of its identical siblings will also appear in the same spot. There's an attraction between them. We will speak, using slightly inexact but vivid language, of an <u>attractive identity force</u> drawing together identical bosons.

Lasers are a spectacular example of bosons (in this case, photons of visible light) in action. You can picture a simple laser as a gas-filled tube, much like a fluorescent light. Gas atoms inside the tube are excited by an electric current, and they return to their low-energy ground state by emitting the excess energy as visible light. What makes the cavity of a laser special is the mirror at each end, which can reflect a photon traveling straight down the tube, bouncing it back and forth repeatedly. Because the photon is a boson, its passage "calls forth" the emission of identical photons from excited gas atoms nearby. A single trapped photon soon becomes an ever-growing surge of identical photons, rushing backward and forward within the cavity in a self-reinforcing avalanche. An intense burst of light—trillions of photons, whose wavelengths and phases are all exactly the same—emerges. Your tube has *lased.*

The constructive interference of bosons is only one side of the story. If we cause two beams of ^{3}He to collide, we find a different surprise: the number of particles in each detector does not stay the same, or double—it drops to zero. It is <u>impossible</u>, in this case, to scatter a particle through 90°. When two identical ^{3}He atoms collide, there is again interference between the alternatives, but this time the interference is <u>destructive</u>. The two possibilities for ^{3}He atoms cancel each other precisely, and at 90° we find no atoms at all.

Particles that behave like ^{3}He atoms are called *fermions,* short for "particles obeying Fermi-Dirac statistics." The behavior of identical fermions is in stark contrast to the behavior of bosons: while bosons like to imitate one another, fermions absolutely refuse to. It is because because two fermions wouldn't be caught dead doing the same thing that there is no scattering at 90° in our experiment. The "identity force" between fermions acts like a repulsion, and the probability of finding a fermion at some point in space is reduced if some of its identical siblings are already nearby.

Fermionic repulsion plays an interesting role in astronomy. When small stars like our sun exhaust their fuel, they begin to contract under the force of their own gravity. Their old means of supporting themselves, by the pressure of hot gas and radiation, fails as heat leaks out and the temperature drops. Eventually, though, a new source of pressure comes into play. Electrons, being fermions, can be forced upon one another only so far. When they are pushed together too closely, repulsive identity forces between them powerfully resist further compression. It is the repulsive identity forces between identical electrons that support white dwarf stars (at a density of many tons per cubic centimeter) against their own gravity. Thus, remarkably, the subtle and semiparadoxical identity force, which couldn't exist if we were able to keep track of which electron is which, becomes the firm support of stars in their last extremity.

Each elementary particle can be judged by its interference behavior and categorized as either a fermion or a boson. Photons are bosons; electrons, protons, and neutrons are fermions.

What about other particles? There is a simple rule for composite objects, such as nuclei (made from protons and neutrons) or atoms (made from nuclei and electrons). The rule is that if such an object contains an odd number of fermions, the composite object is a fermion. Otherwise, it is a boson. Note that this simple rule doesn't care at all about the number of bosons in a composite object. Thus our ^{3}He atom is a fermion because it is made from five fermions—two protons, one neutron, two electrons. Adding one more neutron to produce ^{4}He gives six fermions altogether; so ^{4}He is a boson.

Are there other possibilities, besides bosons and fermions? This question bugged me (Frank) for a long time. It seemed a pity to have only these two extreme possibilities. In fact, it turns out that only bosons and fermions are possible in three or more dimensions but that, in two dimensions, particles come in a continuous range of theoretical possibilities, which slide smoothly from bosons to fermions. If we do the colliding-beam experiment in two dimensions, it is consistent to have constructive interference, complete destructive interference, or anything in between, for different imaginary particles that move only in two dimensions. This was fun to work out, but seemed a little silly, even to its author—after all, the world we know about has at least three dimensions. What joy when, a few months later, people found that these possibilities actually do occur in the world—in real laboratories! Wait a minute—how can that be?

Don't we live in a three-dimensional world, at least? But there are physical systems, in this case, a layer of electrons in a strong magnetic field, which have motions effectively limited to two dimensions.

Isn't everything made from building blocks like electrons, neutrons, and protons? So it seems we need only appeal to the odd-even rule, to find out whether an object is itself a boson or a fermion; and nothing else can occur. But the funny objects don't contain a whole number of electrons; they might contain two-thirds or seven-fifths of one. How that happens is yet another story.

Although it was fun to find these curiosities, the main point remains that under ordinary circumstances everything is either a boson or a fermion. And the fact that nature recognizes exact identity, and puts it to use, underwrites our program of describing the world in terms of an enormous number of exact replicas of a few building blocks.

Building two duplicate towers of LEGO blocks is, as they say, child's play. There's more to creating identical structures, however, than just starting out with a set of identical parts. Much of the cost of any high-performance machine, from a Mercedes to a 747, goes into assuring "quality control"—that is, into making sure that parts that are supposed to be identical really are and to seeing that their placement precisely matches the specifications.

So it remains a puzzle, even given that nuclei and electrons are precisely identical, to understand how nature arranges them into radically identical—indistinguishable—atoms. And similar puzzles recur in other forms at smaller distances: Why do protons and neutrons make radically identical nuclei? Why do quarks make radically identical protons and neutrons? Perfect quality control at every level—how does nature do it?

The popular picture of the atom, a little nucleus inside a whirl of orbiting electrons, promotes this puzzle to true paradox. If atoms really were like miniature solar systems, or like planets surrounded by orbiting moons, they surely would vary in detail from one to another. Each time the electrons got disturbed (that is, each time one atom collided with another), they would take up slightly different orbits. Worse yet, this kind of caricature atom doesn't even resemble itself from moment to moment. As the electrons move along in their orbits, the overall arrangement changes.

Obviously, we need very different ideas to understand the radical identity of atoms—or of molecules, nuclei, protons, and so on. The problem is a general one. Each of these micro-objects is assembled from smaller parts and yet comes out with a definite, unique structure. In fact, the assembled objects reveal their radical identity by behaving like bosons or fermions in the deepest tests we know for indistinguishability.

Amazingly, the physics behind radical identity is closely related to the physics of musical instruments.

Consider the Ur-piano: that is, a taut string nailed down at both ends. It can produce several different pure tones, each one corresponding to a different possible *mode of periodic motion,* or vibration pattern. The geometry of the first three modes, shown in figure 16.3, is simple: almost as if the sounding string can divide itself up into fractions, a series of shorter sounding strings.

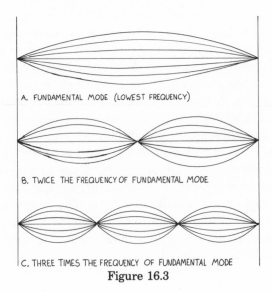

A. FUNDAMENTAL MODE (LOWEST FREQUENCY)

B. TWICE THE FREQUENCY OF FUNDAMENTAL MODE

C. THREE TIMES THE FREQUENCY OF FUNDAMENTAL MODE

Figure 16.3

The lowest frequency, or deepest tone, is emitted by a string vibrating in its "fundamental" mode, mode *a* in the figure. The next-higher tone is emitted when the string vibrates in mode *b*. As is suggested by the geometry, the tone emitted when the string vibrates in mode *b* has half the wavelength (and therefore twice the frequency) of the tone from mode *a*. The tone from mode *c* has one-third the wavelength (three times the frequency), and so forth.

In short, for a string all possible modes of vibration give tones whose frequency is a whole number times the "fundamental" (that is, the frequency of the fundamental mode, our mode *a*). This fact is crucial for getting decent music out. You see, when the string is plucked, several modes of vibration are generally excited at the same time. So the sound emitted is not a pure tone but a chord. By using padded hammers and choosing carefully where to strike, we can make sure to excite more of the fundamental tone than any other, but it is neither possible nor musically desirable to exclude the other tones completely.

The great advantage of having all the frequencies be whole numbers times the fundamental is that the resulting chord is harmonious. The doubled frequency gives a musical octave (C above C, say), the tripled frequency gives the next dominant (the next G), and so forth. Indeed, our sense of harmony is that two tones sound harmonious when there are simple numerical ratios between their frequencies, as Pythagoras discovered. So plucking a string will give a pleasant chord. The difference between the "rich" tones of a good piano or violin and the "thin" pure tones of simple electronic music generators is that the former are actually chords containing a nice mix of higher harmonics, while the latter are strictly single-frequency sound waves. More sophisticated music synthesizers can imitate the sound of a violin or trumpet by duplicating the instrument's characteristic mix of harmonics.

A string is a one-dimensional vibrator. Two-dimensional objects, such as plates, can also vibrate and generate tones. The possible modes of periodic vibration are more varied and complicated in this case. As a result, there is not such a simple relation between the frequencies of the possible tones. They are not harmonious. When the plate is struck, several modes get excited, just as for the string. But this time the resulting chord is a discord. The phenomenon is described in a pungent passage from Helmholtz's *The Sensation of Tone:*

[F]or really artistic music, such instruments as these have always been rejected, as they ought to be, for the inharmonic secondary tones, although they rapidly die away, always disturb the harmony most unpleasantly. . . . A very striking example of this was furnished by a company of bell-ringers, said to be Scotch, that lately travelled about Germany, and performed all kinds of musical pieces, some of which had an artistic character. The accuracy and skill of the performance was undeniable, but the musical effect was detestable. . . .

For similar reasons, the music of steel drum bands must be considered an acquired taste.

The study of two-dimensional vibrators, although it does not turn up any promising new musical instruments, does illuminate the idea of modes. We can extend this idea from the simple vibrations of strings to the more complex vibrations of plates and other objects. Vibrational modes, in fact, are the key to atomic "quality control."

We have learned that electrons are described by laves, that is, by mathematical waves that express the probability for finding a particle. The lave describing an electron in an atom is confined to a region surrounding the nucleus. The electron lave is, of course, neither nailed down like a string nor solidly bordered like a plate; but it is limited by its tendency to snuggle up to an electrically attractive nucleus. The confined electron lave will, like the waves on a string or plate, have various possible modes of periodic vibration. Each mode of the electron lave represents a discrete possible state of the atom. The equations of quantum mechanics express a simple relation between the frequency of a mode and the energy of the state it describes. Just as for photons, the energy of an electron's state is proportional to the frequency of the mode.

Now we are approaching the heart of the matter, the secret of quality control at the quantum level. The key to this extraordinary quality control is simply that there is a finite difference between the frequency of the lowest, fundamental, mode and that of any other mode. Or, translating this statement into the language of energy: there is a gap between the energy of the electron in its lowest-possible mode and its energy in any other mode. The crucial consequence of this gap: if an electron is in its lowest mode, you cannot change its state at all unless you supply enough energy to take it to the next mode. Little collisions do not make little changes; they make no change at all.

If you think about it, you will realize that this last sentence is just another way of saying that atoms are perfectly rigid, and effectively conceal their inner workings. As long as you don't hit them too hard, they behave as if they had no movable internal structure. So even though atoms are made from smaller objects, and fuzzy quantal objects at that, their inner "works" are hidden. The atoms seem (if you don't hit too hard) to exemplify the traditional ideal described by Newton: "solid, massy, hard, impenetrable . . . even so very hard, as never to wear or break in pieces . . . [so] that they may compose Bodies of one and the same Nature and Texture in all Ages. . . ."

What we have said for atoms goes for all the other denizens of the quantal microworld—molecules, nuclei, neutrons, and protons—for the same reasons. All are made from parts that are laves, and laves confined to a finite region come in discrete modes. There is, in each case, a gap between the lowest-possible energy and the next; and we cannot sense the internal structure at all unless we somehow supply enough energy to cross the gap.

This sort of behavior applies to large objects too. Usually, its most dramatic consequences are hidden, because there is plenty of energy (in the form of heat) to cross energy gaps. But at very low temperatures the quantal rigidity of matter produces some wonderfully surprising effects.

For instance, one of the consequences of the quantal rigidity of matter is that at very low temperatures a block of silicon becomes almost free of random vibrations. It takes a certain minimum amount of energy to make the silicon atoms jiggle at all, and at low temperature energy is in scarce supply. The random vibrations are said to be *frozen out*. Left to itself, the low-temperature solid is extremely still. Physicists are now trying to exploit this stillness in a remarkable way. Whenever a particle—a neutrino from a nuclear reactor, say, or an X ray—impinges on a block of silicon and scatters, it leaves behind some energy. The amount of energy deposited by a single neutrino or X ray is minuscule by ordinary standards. Trying to detect these deposits in a room-temperature solid would be like listening for the impact of a pin dropped into a raging surf. But if the block is held well within a degree of absolute zero, its internal noise drops to the point that the impact of single particles can be sensed. Large, sensitive particle detectors based on this principle are being built. Such detectors will have many uses, both in pure science and in technology. They will make it possible, for example, to sense what is going on inside a nuclear reactor, by monitoring its output of neutrinos. Quantum mechanics, by freezing out competing noise, makes audible the faint rustling passage of the neutrino wind.

Now we return to spectra, at last fully prepared to appreciate their message. Light is emitted (in photon lumps) when an electron lave switches from one mode of vibration to another. The emitted photon's energy, visibly encoded in its color, reflects precisely the energy released in the transition between modes. Just as the tones of an Ur-piano or of a steel drum reflect the frequencies of their modes, so the spectrum of an atom encodes the energy of its modes. When we spoke earlier of atoms singing in light, we did not wander far from the sober scientific facts.

(We must admit, though, that since atoms are complicated three-dimensional vibrators, their spectra correspond to horrendous discords. Worse, perhaps, than bell ringers said to be Scotch. It may be all for the best that this particular "music of the spheres" isn't audible.)

The fact that composite microstructures hide their works—in other words, that they can appear perfectly rigid until you bang them hard enough—is a precious gift when we go about using our knowledge, but a curse when we try to augment it. It is a gift for use, because it means that we can predict how a microstructure will behave without having to worry about its internal works. We can do chemistry, or make silicon chips, without ever worrying in the least about what's going on deep inside each and every atom. On the other hand, if we want to know what an electron is made of—or whether, indeed, it has parts at all—we must live with the fact that an electron's behavior below an energy gap of unknown size gives no sign whatsoever of any internal structure.

17

Fields

From a long view of the history of mankind—seen from, say, ten thousand years from now—there can be little doubt that the most significant event of the 19th century will be judged as Maxwell's discovery of the laws of electrodynamics. The American Civil War will pale into provincial insignificance in comparison with this important scientific event of the same decade.
—R. P. Feynman, *Lectures on Physics* (1964)

The reader may well take a skeptical view of Feynman's statement. Even if we agree to pass over events that loomed so large to contemporary men—reforms and revolutions, famines and resettlements—isn't Feynman's choice still unnecessarily obscure? We hope to persuade you that James Clerk Maxwell's work on electrodynamics has done as much to transform our world, and our understanding of it, as any of these events—and its influence continues to grow.

Maxwell's achievement is best understood in its historical context. For centuries, it had been known that certain unusual bits of matter could exert forces that acted on other bodies at a distance. The ancient Greeks, among others, described the peculiar behaviors of electrified matter (specifically, amber rubbed with fur) and of the original magnets (lodestone, found near the town of Magnesia). Each appeared to be a curious power of certain minerals, with no obvious connection to anything else. Nor did electricity and magnetism seem related to one another: the scraps of metal attracted by a lodestone were indifferent to charged amber; the bits of chaff that stuck to amber were unaffected by a magnet. Until the nineteenth century, the study of electrostatics (static electricity is what causes a balloon rubbed with wool to stick to a wall) and magnetostatics (the kind of magnetism found in toy magnets) followed parallel but separate lines.

These studies were shaped by the overwhelming intellectual influence of Newton's theory of gravity. It seemed only natural to use the concepts that had been so successful in describing the gravitational interactions among planets as a model for the interactions between electrified or magnetic bodies. People assumed that there were new forces, electric and magnetic forces, that acted instantaneously across an intervening empty space. And, up to a point, this model works remarkably well. In fact, electric forces are almost embarrassingly like gravity—directed along the line between the bodies, and decreasing with distance as the inverse square of the bodies' separation. The greater complexity of magnetic forces was one small hint that the ideas behind Newton's description of gravity might not be a universal panacea, but the leading mathematical physicists of the time naturally preferred adding epicycles to the existing framework to starting from scratch.

Michael Faraday (1791–1867) was not among those leading mathematical physicists. He had no conventional scientific training at all. Starting as an apprentice bookbinder, he became intrigued with some of the physics and chemistry books he was binding and determined to devote himself to natural philosophy. Faraday educated himself by repeating all existing experiments on electricity and magnetism, then began to do experiments of his own. His research was guided by the intuitive pictures that presented themselves to his imagination in the course of this direct commerce with nature, rather than by contemporary theoretical ideas.

Looking at the patterns formed by iron filings under the influence of a bar magnet (figure 17.1), Faraday had a vision that was to inspire Maxwell, and through him to transform our understanding of the universe. The visible lines of iron filings suggested to Faraday the guiding presence of invisible lines of magnetic force. To Faraday, "empty space" thereby revealed itself as a cornucopia of invisible but powerfully effective structure. Maxwell later contrasted Faraday's vision with the conventional view of his time as follows:

> Faraday, in his mind's eye, saw lines of force traversing all space where the mathematicians saw centres of force attracting at a distance; Faraday

saw a medium where they saw nothing but distance; Faraday sought the seat of the phenomena in real actions going on in the medium, they were satisfied that they had found it in a power of action at a distance. . . .

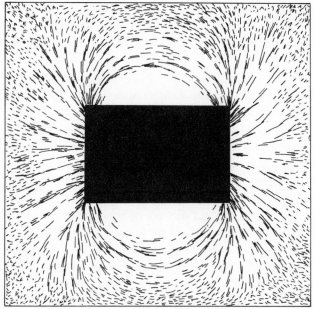

Figure 17.1

If Faraday's point of view still seems obscure, consider this fantasy history of discovery.

Imagine we were able to see fish but were not aware of the ocean. We would notice how when one fish swam by another, the second would swerve a little. We might say that the first exerted a force on the second, which deflected it. But then suppose one day we observed that <u>all</u> the fish we were watching were deflected in a regular pattern, which could not be explained by the forces they exert on one another. Some guy—the Faraday of fish motion—might suggest that the fish are embedded in a medium, which affects them all. And then another genius—call him Maxwell—might suggest equations to describe the medium, and show that it supported self-sustaining disturbances, or *waves*. (Here, for once, we use the word *waves* in its most common sense—ripples in water.) Finally, another clever fellow (Heinrich Hertz—we'll get to him later) would succeed in making waves himself, so that fish far away would swerve at his command. In this way, we would be learning about the ocean, and it would come to take on more and more reality for us. So it was for electromagnetic fields.

Most contemporary theorists took little notice of Faraday's unconventional viewpoint. After all, if one particle acts on a second particle, it seems gratuitous to postulate that the first produces a line of force, which then acts on the second. Thinking about the line of force doesn't seem to add anything. But Maxwell, when he began to work on electromagnetism, made a conscious decision not to learn the advanced theory of his time before studying Faraday's experiments directly: "Before I began the study of electricity I resolved to read no mathematics till I had first read through Faraday's *Experimental Researches on Electricity.*"

It soon became apparent that Faraday's vision of electric and magnetic lines of force filling space had great advantages. In an ironic throwback to the old picture of space as mere stage setting, these patterns of forces came to be known as electric and magnetic *fields,* from the heraldic use of the word *field,* meaning the background setting off a figure. In Faraday's hands, the fields seemed almost to come to life, suggesting new experiments and new ideas that are clumsy even to state in terms of forces between particles.

Faraday's greatest discovery, *magnetic induction,* is a good example. Magnetic induction is a scientific shorthand way of saying that changing magnetic fields produce electric fields. This deep connection between electricity and magnetism had been obscured for earlier observers because the objects of their studies were static—unmoving. Therefore, the fields they produced were static and unchanging. But if you start to think like Faraday, you may get very curious about what happens when field lines shake, or get tangled up. So take a powerful magnet, and set it in motion: the magnetic fields it creates nearby will also undergo a series of rapid changes. Their change creates electric fields. Charged bodies that ignore a stationary magnet fly along the lines of electric force created by a moving one.

Simple examples and analogies are a good way to introduce new physical ideas. Sometimes, as in discussing the bases of quantum mechanics, finding them is quite a challenge. In the case of magnetic induction, however, it is easy to give examples, because this effect is the very foundation of electric power transmission, on which modern life depends. It is magnetic induction that brings the power to your wall plug.

At some distant power station, a waterfall (or a steam turbine driven by a burner or by a nuclear reactor) spins gigantic magnets. Their changing magnetic fields create new electric fields. These electric fields in turn set huge currents of electrons in motion, forcing them through gigantic cables. Through the cables flows the power of the waterfall, eventually divided among many smaller cables for the benefit of users many miles away.

Faraday's humble origins and lack of mathematical sophistication served to obscure the worth of his ideas until James Clerk Maxwell (1831–1879) intervened. Maxwell had all the proper credentials— descent from a wealthy old Scottish family, a degree from Cambridge University, and a reputation as the top mathematical physicist of his era. He also had, like Faraday, a hearty disregard for convention. At Cambridge, Maxwell had made himself notorious for his self-invented "rational" schedule, interrupting his nighttime sleep with a study period from 10:00 P.M. to 2:00 A.M., immediately followed by a bout of exercise—running a circuit throughout the corridors of his dormitory for a half an hour. Fellow students expressed their opinion of this novelty by pelting him with boots and brushes as he passed.

Inspired by Faraday's work, Maxwell set about trying to codify its results into a set of precise mathematical laws. He adopted Faraday's visionary intuition that *fields* rather than particles were of primary importance, and tried to find equations that described how these fields were determined and changed in time. When he did so, he discovered that he could get a consistent set of equations only by introducing a new physical effect, hitherto unknown. The new effect predicted by Maxwell is a sort of converse to Faraday's magnetic induction: that changing electric fields produce magnetic fields.

When Maxwell first theoretically "discovered" his effect, and for several years afterward, there was no experimental evidence to back it up. The effect was difficult to test, simply because the predicted magnetic fields are small and difficult to measure. (Or rather, they were difficult to measure in the late nineteenth century. Nowadays that's duck soup.) Nevertheless, he had enough faith in his ideas to work out their consequences carefully.

Soon thereafter came the revelation: when changing electric fields produce magnetic fields, and changing magnetic fields produce electric fields, a self-perpetuating cycle can occur. A disturbance in the electric field produces a magnetic field—as the magnetic field changes, it produces a new disturbance in the electric field—and on and on it goes, electric-magnetic-electric-magnetic-. . . . *A distur-bance in the fields can take on a life of its own*. Once the process gets started, it does not need help from the outside—from "particles"—to keep going.

Maxwell realized that his equations for electricity and magnetism meant that electric and magnetic fields could form a self-reinforcing disturbance, similar to a wave of sound, which might travel indefinitely far from its source. The marvelous thing was that he could calculate the speed of these electromagnetic disturbances—he found that they traveled with the speed of light. This could not be mere coincidence. Maxwell had no hesitation in identifying his disturbances with light.

Suddenly, a ray of light was revealed as a traveling bundle of electric and magnetic fields. Maxwell saw light itself in a new way. It ceased to be a thing apart and began to be woven into the fabric of physics. And it became clear that visible light is just a special case of a much wider class of disturbances, that visible light consists of electromagnetic waves that happen to have wavelengths in one particular, small range. So it became possible to imagine that there were new windows on the world, new senses as powerful as vision, waiting to be opened. This is the core of the synthesis Maxwell put together in 1860. In his paper "A Dynamical Theory of the Electromagnetic Field," his excitement shines through the conventions of scientific prose (note the three *electromagnetics*):

> [T]he velocity is so nearly that of light, that it seems we have strong reason to conclude that light itself (including radiant heat, and other radiations if any) is an electromagnetic disturbance in the form of waves propagated through the electromagnetic field according to electromagnetic laws.

Maxwell's new view of light was fruitful from the start, bringing together many scattered facts of optics and generating new predictions. But direct evidence for "electromagnetic disturbance in the form of waves" proved more elusive. The problem was that no one was able set up a disturbance in the fields at one place and observe it elsewhere after it had traveled a substantial distance through space. (This is not to say that people couldn't strike a match to produce ordinary visible light, electromagnetic radiation given off by rearrangements in atomic structures. The necessary test of Maxwell's theory was to produce electromagnetic fields—invisible light, if you will—by rearranging larger and more tangible objects in a controlled way.) This feat was first achieved by Heinrich Hertz in 1888, a quarter of a century after Maxwell had predicted its possibility. What Hertz constructed, in essence, was the first primitive radio. Modern radio, television, satellite communication—all are erected upon the foundation laid by Faraday, Maxwell, and Hertz. Their work opened new channels of communication, channels with speed and range far beyond anything available before.

Feynman's assessment of Maxwell's synthesis may well have been prompted by the enormous practical impact of these inventions. We suspect, however, that Feynman was just as impressed by its scientific and philosophical impact—how it changed our understanding of the world's structure. Two of its leading ideas, *field* and *unification,* have been especially fruitful and inspiring.

The essence of the field idea is that the fundamental ingredients we need to describe the world are fields that fill all space. We compared it above to an invisible ocean. More generally, we might consider a liquid mixture of several different chemicals. To describe the mixture fully, we need to know the density of each chemical at every point inside. We might say that each chemical fluid makes its own ocean and that all these oceans inhabit the same space—or that each chemical defines a field filling the liquid. Going back to electromagnetism, we can think of the electric and magnetic fields as fluids filling all space or, as they used to be called, *ethers.*

This idea that space is everywhere permeated with fields is to be contrasted with the older paradigm of point particles moving through truly empty space, after the ideal of the sun and the planets of the solar system governed by Newton's law of gravity. The field picture describes a much richer world, because in the field world there are potential actors at every point of space—not just at those points where there happen to be particles.

We said a much richer world, but we might with equal justice have said a more complex one. The question is, Is this complication really necessary? Why can't we get by with "atoms and the void"?

At one level, this question has already been answered by Faraday-Maxwell-Hertz—fields that can take on a self-regenerating life of their own must surely be included as independent entries in the inventory of reality. But we can and should try to understand this brute fact. Why fields?

Here's the secret, inner meaning of fields: in a world where influence travels at a finite speed, a field embodies the past. To clarify the meaning of this cryptic oracle, let us return once again to fish motion. We have discussed how, by watching fish move, we could be led to discover the ocean. Indeed, we said, sometimes waves go by and all the fish get deflected together in a way that can't be accounted for in terms of forces they exert on one another. But suppose some reactionary Newtonian came up with this one: "Look, you don't need to talk about a mystical ocean none of us can see. I can explain it all in terms of fish. The waves you talk about were actually caused by a faraway giant squid that went into convulsions a while ago. You see, it takes a while for the forces between fish to work. What you thought was the effect of waves was really just a delayed reaction to the squid's convulsions."

What do we say to this? The honest answer is that the reactionary may be right in principle. On the other hand, his point of view is surely obtuse in practice. You see, if the forces between fish are delayed, then in order to describe the total deflection of fish by other fish, you will have to take into account not only where they are and what they are doing now, but also what they were doing all through the past. It gets to be a terribly unwieldy description—if it is possible at all. Instead, we can keep track of everything by saying that disturbances travel through the ocean. In this description, the fish respond only to the present state of the ocean. So we can forget about what the hypothetical faraway giant squid was doing long ago, and just concentrate on how present disturbances propagate through the ocean—a very useful simplification.

Disturbances in the electromagnetic field describe light, and light is laves. Putting these ideas together, we come to appreciate the field in a new way. The electromagnetic field is a medium that, when excited, produces photons. In fact, physicists nowadays often call it the photon field. Just as an ocean wave is a secondary manifestation of the ocean itself, photons are, from our new point of view, second-ary manifestations of a more primary reality—the field. And now we begin to understand the precise identity of indistinguishable photons. They are all flows in the same ocean, so to speak, merely set in motion at different times and places.

What we have said for photons goes for electrons, or other "particles," too. Each is described by its own field, a medium filling all space that, when excited, gives birth to the particle. The field description is necessary not only for photon laves but also for electron laves. With this, we have traced the amazing indistinguishability of the building blocks of nature to its root. We understand that it is a symbol, reminding us that each of these building blocks is but a disturbance in one of the fields, a wave in one of the several unseen oceans filling all space.

The Newtonian picture of a world populated by many, many particles, each with an independent existence, has been replaced by the field picture of a world permeated with a few active media. We live amid many interpenetrating fields—one for photons, one for electrons, and so forth—each filling space. The laws of motion, in field language, are rules for flows in this ocean. And the rules of transformation—describing the weak interaction or the emission of light, for example—are, in this picture, telling us what chemical reactions occur among the components of the universal ocean.

We have compared the universe, as described by field theory, to interpenetrating oceans. Another metaphor, though perhaps less familiar, is in some ways more fitting.

A major component of many stringed instruments is a *sounding board,* which resonates with the sounds produced by plucking the strings, and thus amplifies and shapes them. The underside of a piano contains a sounding board, and the bodies of violins or guitars play the same role.

Let us imagine a <u>universal</u> sounding board filling all space. The universal sounding board can be excited at different places. Whenever and wherever it is excited, it starts to vibrate in the same way. Vibrations initiated in a small region can spread at a finite speed through the sounding board. In all this, vibrations in the sounding board resemble excitations in the fields filling space.

Material particles, then, are vibrations in the universal sounding board. The primary structure is the sounding board; what we ordinarily call matter is a secondary manifestation. The possible forms of matter, and their interactions, are implicit in every small cube of apparently empty space.

We must imagine our sounding board as ever humming, for it is a quantum-mechanical object, and like all such it incessantly jitters. It hums, and this hum, interpreted in particle language, is the sound of virtual particles coming to be and passing away.

Dents in the sounding board are not precisely localized at a point, but taper off gradually. Thus the presence of a particle alters the fields nearby and affects the motion of other nearby particles. And if the forces making a dent are suddenly relieved (a particle decays), the sudden release of energy sets vibrations in motion (the decay products move off). The elastic response of the universal sounding board illustrates the dual roles of fields, which are both transmitters of forces and creators of particles.

Maxwell's synthesis had yet another great consequence for the scientific view of the world. It has served not only as a source of fruitful concepts—most important, of the field concept—but also as an inspiring example of the possibility of *unification*. It completed the process of unifying electricity and magnetism into a single, close-knit larger subject called, of course, electromagnetism, and it showed that optics—the study of light—is also subsumed in it. As an example of how utterly diverse phenomena (what do rubbing amber with cat's fur, minerals that align themselves north and south, and light have in common?) can be shown to have a deep common origin, this is hard to beat.

Maxwell Redux

R. P. Feynman's admiration for Maxwell did not stop him from tinkering with Maxwell's legacy. Far from it! In fact, he created a new form of physical language, called *Feynman diagrams,* which reproduced Maxwell's equations in a form Maxwell himself would scarcely recognize:

Maxwell's equations, old and new.

Old	New
$\nabla \cdot E = \rho$	
$c\nabla \times E = -\dfrac{\partial B}{\partial t}$	
$\nabla \cdot B = 0$	$\mathcal{W}\mathcal{W} =$ photon
$c\nabla \times B = j + \dfrac{\partial E}{\partial t}$	$\uparrow =$ electron

In the left-hand column, you see four equations, essentially those put together by Maxwell, whose esoteric mathematical appearance should not disguise the enormous simplification they represent. In the right-hand column, a silly-looking doodle of a photon bumping an electron reproduces the same physics (and more) in today's notation.

This doodle is an example of a Feynman diagram. If you wander into the office of any practicing particle physicist, chances are you'll see several such diagrams scrawled on the blackboard. What do they mean?

A Feynman diagram might be described as the ultimate in bird's-eye views, offering an overview of disturbances in the fields as they exist not only throughout space but also in time. The vertical distance between two events shows their separation in time; the horizontal distance shows their separation in space. There are mathematical expressions associated with each diagram, which tell how likely it is that the space-time process the diagram depicts will occur. The expressions, it turns out, are amazingly simple.

As an example, consider the basic "Maxwell-Feynman diagram":

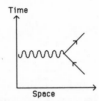

In this diagram, the disturbance in the photon field has no movement in time. It might represent a fixed electric or magnetic field. The disturbance in the electron field is moving in time—it's a fancy way of drawing an ordinary electron. So this diagram shows the deflection of an electron by a fixed field.

The basic Feynman diagram describes a "chemical reaction" between the electron field and photon fields—the process that couples the photon field and the electron field. In the variations on this basic coupling, all the interactions of electrodynamics are implicit. (Strictly speaking, we must include fields for other charged particles—protons or quarks—and their couplings to the photon field too.) Consider, for example, the Feynman diagram describing the interaction between two electrons, shown in the following figure:

This figure shows how ordinary electric repulsion between electrons arises from the coupling of their fields. The presence of one electron creates a continuing disturbance in the photon field, which acts in turn to repel the other electron. Notice that in this picture the disturbance in the photon field does not pass beyond the interacting electrons—no light gets emitted into space. But it is convenient and usual to say that *virtual photons* are being passed between the electrons.

The profound point here is that the existence of forces and the existence of particles have become two sides of the same coin. When the disturbance in a field is short-lived and confined to a small region of space, we say it is a *virtual particle*. The exchange of virtual particles generates a force between the particles doing the exchanging, such as the repulsion of electrons in our present example. When the disturbance in a field persists for a long time and travels a long way, we recognize it as an ordinary or *real particle*. In a field picture, real and virtual particles, or particles and forces, are just two possible kinds of excitations of the same field.

The shifted viewpoint of the Feynman-Maxwell diagram, in all its variations, can be used to unify and simplify our understanding of many other physical process. Consider the two shown in the figure below:

a   b

The first part of the figure *(a)* shows a photon—a real one, this time—traveling along, getting absorbed in one coupling and emitted in another. The net effect is that it has bounced off an electron—interacted and changed direction. In the jargon, we say the photon has *scattered*. Many familiar processes with light—reflection in a mirror, the working of lenses and microscopes, the rainbow, the blue of the sky—are consequences of such scattering.

The second part of the figure *(b)* shows something a bit more exotic. A photon appears from somewhere and disintegrates into two electrons. But wait—one of the electrons is traveling in the wrong direction, backward in time. Actually, the process pictured here is the disintegration of a photon into an electron and an antielectron. An electron moving backward in time represents an antielectron, or positron. We are playing here with ideas we will fully explain in chapter 18; but we wanted to show that the innocent-looking little coupling of fields contains implicitly not only Maxwell's equations but also some processes Maxwell never dreamed of.

What do you see in the figure above? Is it a white candlestick on a field of gray or two dark silhouettes on a field of white? Such drawings are often called optical illusions, but in fact they present a conceptual choice: to decide what's the figure and what's the background.

Choices of this kind have filled the two preceding chapters, beginning with (and including) Feynman's perception of the most important event of the nineteenth century. It is a perception we share. Just as Bach's music now completely eclipses any memory of the wealth and prestige of his patrons, or the wars of petty princes that surrounded him; just as one can still, after millennia, read the work of Archimedes with fascination and pride in being human while the much-admired exploits of contemporary conquerors evoke only pity for their futile vanity; so Maxwell's achievement will loom larger and larger to future generations who continue to reap its fruit, and the squabbles that now fill history books will come to seem sad curiosities.

The vision of Faraday that set the stage for Maxwell was likewise a figure/ground decision. Whereas earlier workers had seen the drama of electricity and magnetism as a play with particles in all the starring roles, Faraday focused on the lines of force in the spaces between them, and Maxwell perceived background fields controlling the drama.

Feynman was to suggest yet another change of viewpoint: from that of watching the orderly evolution of fields forward through time to that of concentrating on the space-time trail left by disturbances in the fields. With each new choice, each viewpoint, new features of reality are brought to light.

In the figure, the candlestick looks reasonably symmetric. When you shift to looking at two profiles, however, you will notice that they are two rather different faces. Alerted to this, you will, the next time you look at the candlestick, no doubt see new features—asymmetries that previously escaped you.

It may appear, from the new form of Maxwell's equations, that his grand synthesis has been reduced to child's play. And, in a sense, it has. But this in no way should be regarded as diminishing his achievement—quite the contrary. Just as pioneers may clear a path through a forest leading on toward a mountaintop, Maxwell and those who followed in his footsteps broke through a thousand hindrances on our behalf. If we follow his path and discover, looking down from the heights he aimed for, that more direct pathways were possible, we still must marvel at the determination and strength that first forced a trail through the wilderness—and remember with affection the enthusiastic young Scottish professor who led the way. Blessed are the trailblazers.

Virtual Particles

Beware of thinking nothing's there—
Remove what you can, take infinite care
There still remains a mindless seething
Of frenzied clones beyond conceiving.

In a wink, they arrive, and dance about,
Whatever they touch seems inspired to doubt:
Shall I change my identity? Move away?
These thoughts often lead to a rapid decay.

Mourn not! The terminology's misleading:
Decay is virtual particle breeding,
And seething, though mindless, may serve noble ends:
Clone stuff, exchanged, is a bond between friends.

To be or not? The choice seems clear enough.
Hamlet vacillated; so does this stuff.

Seventh Theme

Transforming Principles

Now we draw bridges between the intangible but fundamental, and the tangible but impermanent, elements of reality. To begin, we describe the process by which energy takes form as matter and antimatter. Then we discuss the modern theory of the strong interaction, quantum chromodynamics or QCD, in depth. In ascribing the forces between quite tangible protons and neutrons to unobservable gluons transforming the unobservable properties of unobservable quarks, and doing so fruitfully and convincingly, QCD provides a striking example of a richly imaginative reality hidden beneath less glamorous appearances.

Here, too, you will find first-person account of the birth of asymptotic freedom, a concept that is the key to understanding the workings of QCD.

Prelude Seven

THE SEARCH FOR DEPTH

The ancient Babylonian priests and scribes who kept the calendar prepared great tables and devised ingenious rules of thumb to describe the changing patterns in the sky. Although they could predict the motion of the planets and even eclipses of the sun and moon, we would not be inclined to say that they had a deep knowledge of astronomy. They did not study the stars and planets as physical objects. The old astrologers missed wonderful discoveries and never began to construct a scientific cosmology. They had collected many answers, but they had not asked enough questions.

If we are to avoid a similar pitfall, we cannot long remain satisfied with the complicated description of the force between protons and neutrons that is inferred from patterns of nuclear stability. To wit, the force is short-range, a strong attraction between particles 10^{-13} centimeters apart that turns into a repulsion when they come just a little closer and decreases sharply at larger distances. The magnitude and direction of the force also depend on the direction of the spin and velocity of the particles involved. To put it bluntly, the force seems about as complicated as it could possibly be, depending significantly on every variable in sight.

(One simple regularity is that the strong force doesn't distinguish between protons and neutrons. There are a few other simplifications based on symmetry, and we shall get to them later. These constraints on complexity, while interesting and important, still leave us with a complicated-looking mess.)

It would be possible, by careful and exhaustive experimentation, to determine the nature of the strong force in all circumstances and to record the results in tables once and for all. From these tables, we might hope to explain quantitatively and in detail the structure of nuclei, and even to predict the existence of forms and phenomena not seen before or difficult to reproduce in terrestrial laboratories. In fact, this sort of activity is the subject of traditional nuclear physics. It has given us, among other things, surprising insights into the history of the universe and the origin of chemical elements, the metabolism and life cycle of stars, and the dynamic geology of Earth and other planets. It has also given us the infinite menace of nuclear weapons and the potential of unlimited cheap energy from fission and fusion reactors. Earlier in this book, we were able to discuss some of these things using very simple facts about the strong force. But this rich variety of applications should not be allowed to obscure the reality that the underlying description of the force lacks intellectual coherence and beauty—in a word, *depth*.

What is depth in a scientific theory? There is no universally accepted definition; indeed, the concept seems rarely to be discussed in print. It is, however, often used informally, and most scientists would surely claim to "know it when they see it."

There is much more to science than the traditional categories of truth and falsity; ideas can be true but trivial (90 percent of what appears in scientific journals) or strictly speaking false but deep and fruitful (for instance, Newton's theory of gravity). Let us try the following definition, inspired by recent work of Charles Bennett and others in computer science. A theory is said to be *deep* when it satisfies two requirements. First, it must have verifiable consequences that can be derived from the premises of the theory only by long chains of logical deduction. Second, there must not be another theory that leads more easily to the same consequences from fewer premises. In less formal language, the first sign of a deep idea is that its consequences are not immediately apparent. Einstein's theory of gravity is in this sense extremely deep, since it is stated in terms of rules for determining the curvature of space-time, which have no obvious connection to the familiar force of gravity at all. Indeed, it takes a good deal of work involving tensor calculus to make the connection. Having hidden consequences is not enough to make an idea deep, though—otherwise we could manufacture deep ideas as easily as anagrams. The second requirement is that the work we have to do to find the consequences is really necessary. If all Einstein's theory of gravity did was reproduce Newton's laws of gravity from a different and more obscure starting point, it would be considered not a deeper theory but rather the same theory written in the form of a puzzle. It is largely because a striking property of gravity—its universality, the fact that all bodies acquire the same motions (accelerations) in a gravitational field—is built into Einstein's theory but must be grafted onto Newton's that all the extra work is worthwhile.

(We do not, of course, mean to deprecate the genius of Newton, whose theory of gravity is itself extremely deep. To derive the motion of planets and tides from the basic inverse square law of gravity is a nice exercise in calculus even today, and it required immense mathematical skill and ingenuity, given the tools available in Newton's time.)

Deep theories are often found to possess a special sort of simplicity, what might be called radical simplicity. In the ideal case, the theory is formulated as a single equation. This single equation, however, will refer not to anything immediately observable but rather to inferred underlying structures. Examples are the geometry of space-time in Einstein's gravity theory or the quark and gluon fields in modern strong-interaction theory. Although these structures are the fundamental elements in their respective theories, they are hard to observe directly.

The behavior of the underlying structures is in a radical sense simple—summed up in the one equation. The observable features of the world, the features that our sense organs and measuring instruments happen to be well adapted to perceive, can, however, be elaborate combinations of these simple ingredients. The observable combinations may behave in complicated ways that require a lot of mathematical and logical work to derive, though ultimately they are consequences of the single equation of our deep theory. If we are to understand complicated appearances in terms of underlying simplicity, this is the price that must be paid.

Depth is not at all the same as fruitfulness. Depth is like the root system that supports the visible tree. A barren tree may have deep roots; inversely, a tree with shallow roots may bear much fruit. Einstein's theory of gravity, at the time it was formulated, made very few testable predictions that could not be derived, much more easily, from Newton's theory. In the long run, however, the deeper roots will support new growth. During the last twenty years, Einstein's theory has become a necessary, workaday tool in the description of extreme astrophysical objects like neutron stars and black holes—and the foundation of big bang cosmology.

The search for depth is, we think, partly what can only be described as an artistic impulse: to make our theories more self-contained, intricate, inevitable—and therefore beautiful. As a bonus, we hope and expect it will yield new fruit, as it always has in the past. A visitor to the gaslit halls of the London Exhibition of 1851 asked Michael Faraday what possible use could arise from his experiments on electricity. Faraday replied with a question of his own: "Of what use is a newborn baby?"

18

Antimatter

Up to this time we have spoken as simple physicists: *now we must advance to* metaphysics *by
making use of the* great principle, *little employed
in general, which teaches that* nothing happens
without a sufficient reason. . . . *This principle laid
down, the first question which should rightly be
asked, would be* Why is there something rather
than nothing? *For nothing is simpler and easier
than something.*

—G. W. Leibniz, *The Principles of Nature
and of Grace* (1714)

Leibniz posed what must surely be the ultimate problem about the
origin of the universe: Why is there something rather than nothing?
We make a promising start on this problem by showing how <u>many</u>
arise from <u>few</u>.

In a typical modern high-energy physics experiment, two rapidly
moving protons are smashed together and hundreds of particles
emerge. A natural thought is that the protons have been broken to
bits and that their parts have come flying out. However, when the
stuff that comes out is analyzed, it becomes clear that this impression is completely wrong and that something more interesting has
occurred. For when we identify the "bits" that have flown out, we
find lots of protons and a whole zoo of other particles, many of them
heavier than the protons we started with. It is clear that the slew
of particles emerging from our collision cannot be considered as
preexisting pieces of the original protons. A new principle of fecundity is at work—nature is producing more from less, particles from
pure energy.

Einstein's equation $\mathbf{E} = \mathbf{mc^2}$, perhaps the only equation so famous that many people can recite it without knowing what it means, embodies the idea that the Energy and mass are really somehow equivalent, or at least interchangeable. (The conversion factor, c^2, is the square of the speed of light.) In the burning of stars or the explosion of nuclear weapons, some of the mass of nuclei is converted into energy. Radiation is emitted and matter is heated and accelerated outward. In our proton collision, the equation can be read the other way—from the energy concentrated in the incoming protons, products of larger total mass are emerging.

Strict rules govern the flow and counterflow between the realms of energy and matter. Einstein's equation tells only a small part of the story and gives little hint of what <u>kind</u> of matter arises out of energy. We shall show in this chapter how the generation of new kinds of particles from energy is an inevitable consequence of the fiery marriage between relativity and quantum mechanics. We will demonstrate that, unavoidably, every particle must have an *antiparticle* with the same mass as the particle but the opposite electric charge. <u>The basic production process, which underlies the fecundity of nature at high energies, is the creation of particle-antiparticle pairs from pure energy.</u>

The first antiparticle to be discovered was the *antielectron,* also called the *positron.* You can appreciate that the raw material studied by physicists does not come with labels like "I am a positron." Only by gathering many clues can we identify the suspects, so to speak. A lot of cleverness goes into teasing out the maximum amount of information from just a few hints provided by nature. When Carl Anderson "discovered the antielectron," he actually observed particle tracks similar to those shown in figure 18.1. Here are some of the clues he had to work with:

1. *The setting.* The picture was taken in a bubble chamber placed inside a large magnetic field. Tracks were left by cosmic-ray particles passing through the chamber. The tracks that need explaining are the curlicues diverging from point X.

2. *What's the charge?* Only electrically charged particles leave any tracks at all in a bubble chamber. A charged particle will spiral to the right or to the left in a magnetic field, depending on whether its charge is positive or negative. So one of the particles emerging from X is positively charged and the other negatively charged. (You see now why it was a clever idea to have the magnetic field.)

3. *What's the mass?* Other things being equal, a heavier particle, since

it has more inertia, will be less bent by a magnetic field. However, the speed or, equivalently, the energy of the particle also comes in. You can get a handle on the energy by studying how far the particle travels before stopping. There is a complicated, but basically foolproof, procedure for putting together the two pieces of readily available information about the track—its curvature and its length—and determining the mass of the particle that caused it.

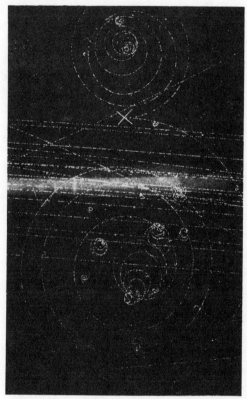

Figure 18.1

After carefully studying several similar pictures, Anderson identified the oppositely curling curlicues as the tracks of oppositely charged particles—one of them the familiar electron, the other a particle never before recorded. The new particle, he found, has the same mass as an electron but the opposite electric charge.

The upshot of this detective work was to identify the event at point X as the simultaneous creation of an electron-positron pair. (From nothing? So it would seem from the picture, but the picture doesn't show everything. Only electrically charged particles leave detectable tracks. Very likely a photon, which is neutral and leaves no track, dumped the energy necessary to create a pair.)

Anderson didn't know it at the time, but his discovery of positrons had been anticipated theoretically. Just by thinking about the implications of relativity and quantum mechanics, the theorist Paul Dirac had been forced to introduce such particles into his equations. Although it is a little more difficult than most of the things we discuss in this book, we cannot resist presenting some of the logic behind this extraordinary theoretical discovery. If you are willing to put some imagination and patience to work, prepare to appreciate one of the most surprising, beautiful, and central hidden harmonies in modern physics.

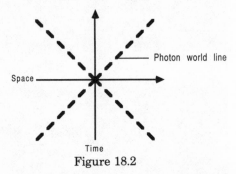

Figure 18.2

Figure 18.2 brings us back to the world overview of the Feynman diagram, where time moves vertically and space is one-dimensional and horizontal. The history of a particle plotted on such a graph is called its *world line*. For example, the world line of a stationary particle would be completely vertical (parallel to the time axis): its time coordinate changes as it keeps on existing, but its position in space is constant. The faster a particle moves through space, the more its world line will be tilted away from the vertical.

Consider the possible world lines of a photon, shown here by dashes. The speed of light in vacuum is a constant of nature. A (real) photon's constant speed means that its world line will always be absolutely straight, with no curves or kinks, and will always lie at the same angle from the time and space axes. In figure 18.2, we've chosen the units of length and time so that possible photon world lines, shown by dashes, lie at a 45° angle, halfway between the two axes.

The theory of relativity teaches us that the speed of light is unusual in another way—it represents a limiting velocity. Nothing can travel faster than light. For our picture, this translates into the statement that the tilt of a particle's world line never exceeds 45° from the vertical.

So far, nothing terribly strange. But now let's consider an electron moving at very close to the speed of light. And let's remember what we have been sweeping under the rug, that we should really be describing the electron as a lave. Its position, in other words, is uncertain; indeed, we must think about exactly how we plan to measure it if we are to assign the position any sure meaning at all. If the position is uncertain and the electron is moving on the average at close to the speed of light, then if we sample it at two times, we must anticipate that sometimes it will appear to have moved <u>faster</u> than the speed of light in the interval. Figure 18.3 makes this crystal clear. (To avoid clutter, we've left out the time and space axes but kept the dashes that stand for motion at the speed of light.) Notice that the speed of light can be exceeded only for short time intervals; the cloudiness in position can't change the average slope much in the long run.

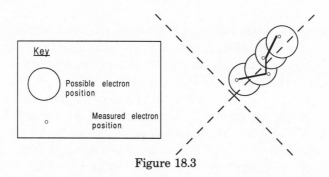

Key

◯ Possible electron position

○ Measured electron position

Figure 18.3

(Wait a minute—doesn't the theory of relativity outlaw such speeds? This apparent violation of the relativistic speed limit gets rationalized in the usual way of quantum paradoxes, by paying close attention to the measuring process. In order to measure the position of the electron, we have to probe it, say, by scattering light from it. So we are disturbing the electrons. Now, the relativistic speed limit strictly speaking applies to the transmission of information: you can't pass information along faster than the speed of light. This limit <u>does</u> remain in force. Suppose someone tried to send you a message by fast electrons. To read the message, you have to disturb the electrons. But by disturbing the electrons, you garble the message. A mathematical analysis shows that you can't win: to measure velocities faster than light, you have to do fast measurements, which greatly disturb the electron.)

So when we do quantum mechanics, we must consider wild world lines, with bigger tilts than light—more than 45° from the vertical. Once you start considering big tilts, there's no end to it. You are driven to consider really wild world lines, where the electron doubles back on itself, as in figure 18.4. We will shortly give you a beautiful but rather subtle logical argument that you <u>must</u> take into account such ultrawild paths. For the moment, let's take it for granted that you do encounter such paths, and try to figure out what they mean.

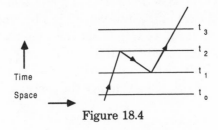

Figure 18.4

Everything is normal up to the time we have labeled t_1. One electron is moving along a normal world line. But at time t_1, clearly something funny has happened. Whereas at first there was one particle, suddenly, starting at t_1, there appear to be three. Looking at the diagram, we notice another thing: two of the particles, the ones on the outside, are moving forward in time as we proceed along the world line—but the third, in the middle, is moving <u>backward in time.</u> The trajectories moving forward in time represent ordinary electrons, but what sense can be made of the backward one?

Taking the proper radically conservative approach, we hold fast for as long as possible to well-tested ideas. We try, for example, to keep believing that the conservation of electric charge holds true. If the initial particle is an electron (charge -1), the total charge of particles at every stage must add up to -1. In particular, the three particles present from t_1 to t_2 should have a total charge adding up to -1. Since each of the two ordinary electrons contributes -1, the oddball must be a particle with charge $+1$. We can argue the same thing another way: since the process at t_1 represents the production of two particles "from nothing," these particles must be able to cancel each other in every respect. In particular, they must have opposite values of such conserved quantities as electric charge. The oddball is appropriately named an antielectron.

R. P. Feynman first developed this intuitive way of understanding antimatter, revealing it to be the inevitable result of the merger between quantum mechanics and relativity. He described his realization in a striking image: "It is as though a bombardier flying low over a road suddenly sees three roads and it is only when two of them come together and disappear again that he realizes that he has simply passed over a long switchback in a single road."

Now let's close the logic of our argument, by showing that the ultrawild paths can't be avoided. We will do this by taking a different view of the slightly wild paths. We will find that a slightly wild path, one containing a faster-than-light segment, appears as an ultrawild path, one containing a backward-in-time segment, to an observer moving very fast in the same direction. Since the basic principle of relativity theory is that any observer whose own velocity is constant sees the same laws of physics, this will show that ultrawild paths must also be included in <u>our</u> description of the world.

As Maxwell taught us, the speed of light is built into the laws of physics. When moving observers cling to the constancy of the speed of light with proper relativistic fervor, however, puzzles arise. It turns out that concepts of time and space themselves get scrambled for observers in motion. Einstein, in his 1905 paper on relativity, reached this conclusion by analyzing a seemingly innocuous problem: How can two people far apart synchronize their watches?

One way is the following. To synchronize your watch with a friend in a distant rocket ship, you bounce a signal beam of light off his rocket. <u>He</u> sets his watch to noon when he receives the signal; <u>you</u> set your "noon" to halfway between the time you sent the signal and the time you got it back. (In practice, you would measure the interval between sending and reception and would at reception set the time to noon plus half the interval.) The result of synchronizing watches among a whole family of stationary observers is shown in figure 18.5. You see that, just as the lines of constant position are vertical, the lines of constant time are horizontal.

The exchange of light signals may not always be the most practical way to synchronize watches. But it does have the fundamental ad-

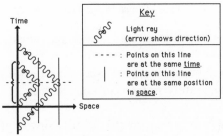

Figure 18.5

vantage that the speed of light is a universal constant, so that this method is unambiguously tied to basic physical laws.

Now let's see how observers on another fleet of rocket ships synchronize their watches. The world lines in figure 18.6 stand for rocket ships moving rapidly to the right, from our perspective. However, because their velocities are equal—their world lines are parallel—they seem to one another to be perfectly motionless. If they were drawing the picture, you would see them set their world lines precisely vertical to reflect this; instead of zipping along through space, they perceive themselves at rest in what we shall refer to as *scape* (mixed-up space).

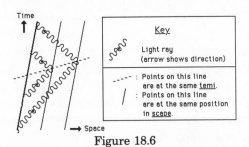

Figure 18.6

They synchronize watches by exchanging light signals, whose world lines are all at 45° from the vertical. Whether in space or in scape, the speed of light is constant. You see from the picture that when we watch those ships follow our prescription for synchronizing, we see them arrive at something different from our time. Events they agree are happening simultaneously—let's say at the same *temi*, * for their mixed-up time—are not happening at the same time, as far as we stationary people are concerned. Or, to put it another

*I (Frank) have been looking for a chance to introduce this word, *temi*, into the English language for a long time. You see, it completes a cycle of four-letter words: *emit–mite–item–temi*. As far as I know, this would be the only such cycle in the language. I hope you will help me make *temi* popular, so that the *Oxford English Dictionary* will eventually legalize it.

way, the lines of equal temi are not horizontal, so they intersect an infinity of horizontal lines—an infinity of times.

You should convince yourself that as we let the speeds of the moving observers increase toward that of light—so the slope of their world lines approaches 45°—the lines of equal temi tilt farther away from the horizontal. In fact, these lines of equal temi approach (but never quite reach) a slope of 45°, as we consider observers moving closer and closer to the speed of light.

Now, finally, look at figure 18.7. You see immediately that a particle moving forward in time can be backtracking in temi. To observers using temi instead of time, a slightly wild path becomes an ultrawild path. This closes the logic of our argument.

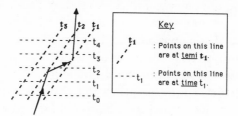

Figure 18.7

As a bonus, you can now appreciate the logic behind the discussion in chapter 8, where we mentioned how neutron decay implies a variety of kindred processes. The basic process, once again, is

$$\mathbf{n} \rightarrow \mathbf{p} + \mathbf{e} + \nu$$

Now, however, you realize it's also possible that the electron follows an ultrawild path. And you know what that means: instead of an electron emitted by the neutron, there is a positron absorbed. In other words, the process becomes

$$\mathbf{n} + \mathbf{\ominus} \rightarrow \mathbf{p} + \nu$$

This ultrawild logic gives rise to the general rule: you can take any particle in a reaction and cross it over to the other side to obtain a possible reaction for the antiparticle.

If the electric charge of the antiparticle is opposite to that of the related particle, what about the mass? We will argue that it must be equal.

Before we do this, though, it's fun to consider for a moment what would happen if antimatter had "antimass." What would a negative mass antiparticle (NMA for short) be like? Contrary to what your first instinct suggests, the NMA would not fall up. The gravitational force is proportional to the mass, but acceleration is force divided by mass; these two effects cancel, so that changing the sign of the mass does not change the motion. On the other hand, the gravitational force exerted by the NMA repels both negative and positive masses.

Similarly, the NMA would respond to an "attractive" force—such as the electric force from its oppositely charged partner—by moving away. So the particle would chase the antiparticle, which would, however, accelerate away in full-scale retreat. This may remind you of certain absurd and unfortunate human relationships, and it is even more absurd as a physical phenomenon. For one thing, it would provide an unlimited, runaway source of energy—a mixed blessing at best, because it makes the world unstable.

Having convinced ourselves that the obvious guess of opposite masses is absurd, we return to understanding why the masses are equal. An argument can be constructed starting from the principle of time-reversal symmetry we met before. This principle says, basically, that the laws of physics are the same if time is run backward. Now, of course, the behavior of a particle depends on its mass, so if an electron running backward in time—a positron—is going to behave the same way as one running forward in time, it had better have the same mass. Thus electrons and positrons must have the same mass, according to the principle of time-reversal symmetry.

If you balk at this argument—congratulations! You are quite right to do so. It proves both nothing and too much. It proves nothing, because it brings in a new assumption about time-reversal symmetry. Surely, in considering a strange new situation like the existence of antimatter, we must be careful about assuming things. On the other hand, it proves too much, because it applies equally with *charge* substituted for *mass* everywhere. But we know that the charges of electron and positron are opposite, not equal.

Fortunately, both these loose ends can be tied up, in an interesting way. We have mentioned how what the moving observer calls temi and scape are mixed-up, "rotated" versions of what we call time and space. Nevertheless, either description is equally good, and he successfully uses exactly the same laws of physics as we do, according to basic tenets of relativity theory. Now, a purely mathematical analysis of the most general space-time "rotations" shows that it is possible to make a sequence of "rotations" that wind up reversing the direction of time. So time-reversal invariance is not really a separate assumption; the possibility of running the world backward in time is already implicit in relativity theory. But the mathematical analysis (to which we cannot do justice here) actually forces us to refine and slightly modify the idea of running time backward. The operation that is guaranteed to work, is this: <u>run time backward, interchange left and right, *and* interchange particles with antiparticles.</u> This combined operation is called CPT (because it involves reversing charge C, reversing the left-right parity P of space, and reversing time T), and it is guaranteed to generate a world with the same laws of physics as ours. Using this modified operation in the same way as we cheatingly used simply T before, we can convince ourselves that the positron must indeed have the same mass as the electron but opposite charge.

Once we have settled that the masses are equal—both are positive—another paradox seems to arise. If both particle and antiparticle have positive mass, it takes energy to produce a pair. How can we reconcile this with the idea that antiparticles appear "from nowhere" in ultrawild paths, as a purely <u>logical</u> consequence of the quantum-mechanical uncertainty in position? Surely, it can't take energy to enforce logic!

Here, as in other cases we have met, quantum mechanics is poised on the knife-edge of paradox. Violent fluctuations in position, such as the kink in an ultrawild path, are very short-lived. Although they involve very rapid motion, they make only a tiny change in the electron's position. Now, to ascertain the position of the electron with high accuracy, we must view it with a photon or some other projectile having enormous energy. Low-energy laves have large wavelengths—in other words, big uncertainties in position—and so we can't use them to get a precise fix on the electron's position. But if the probing lave has a large amount of energy, the probe itself is capable of providing the energy necessary to produce the pair. In other words, to *see* the kink in an ultrawild path, we must make available the energy necessary to *produce* it. In this typically tricky way, quantum mechanics maintains its logical self-consistency.

The same arguments that gave us the antielectron lead us to expect that there must be antiprotons, antineutrons, antiwhatever. And sure enough, such antithings do get produced, as soon as enough energy is available. In fact, these antiparticles are produced so copiously at high-energy accelerators that we may begin to wonder, What's so "anti" about them? Or, to put it more plainly, Why is the world made of protons, neutrons, and electrons instead of antiprotons, antineutrons, and positrons? It's a real puzzle, because there is an exact symmetry of physical laws—CPT, of recent note—which interchanges matter and antimatter. We shall return to this puzzle, and find ourselves grasping its likely solution, in chapter 27.

Now we can begin to frame our answer to Leibniz's great question. We have found a hint that our rich and varied world can arise, not quite from nothing, but from just one very simple and definite thing—namely, energy. For, given energy alone, we have seen how nature in her fecundity will give us material—particles and antiparticles—that, rearranged and combined in complicated ways, can certainly generate a rich and varied world. A program to attack the ultimate problem of our origin therefore takes definite shape: let us suppose that energy is given, and see if its products congeal into something resembling the world we observe. This grand program is the subject of chapter 27; there we shall trace recent, encouraging progress toward carrying it out.

19

Quarks: A Peculiar Chemistry

Here be dragons.
                                        —E. Fermi
(After a lecture describing the structure of atoms,
Fermi drew a circle around the nucleus and made
this comment, echoing ancient mapmakers.)

The forces between protons and neutrons, which hold atomic nuclei together, seem terribly complicated when we look "from the outside in." At the outer limit of their reach, at distances from 10^{-13} centimeters, to 10^{-14} these forces depend on distance, velocity, and spin in a thoroughly tangled way. The only hope for finding a radically simple and deep description is to work "from the inside out." In other words, we hope that by looking inside the protons and neutrons, we will find what makes them tick.

Such a strategy worked spectacularly well for atoms. Interactions between atoms can be extremely complicated, of course—all chemistry bears testament to this. Nevertheless, working "from the inside out," we find that all these complications arise from radically simple laws—the laws of electricity and magnetism, in the framework of quantum mechanics—acting among simple, identical electrons.

To get a look inside protons or neutrons, one must probe them with high-energy projectiles. It is only with very energetic particles that small regions of space, such as the interior of protons, can be explored. The laves associated with low-energy probing particles are spread out, and the information they can produce is only an average for a similarly spread-out volume of space. Just as you cannot discern the shape of headlights in a fog, so you cannot discern the structure of protons from their effect on diffuse laves.

Motivated by such considerations, experimentalists built big accelerators that could produce beams of high-energy protons—or other particles, such as electrons and photons—and studied what happens when they smack into things.

If speedy simplification was the goal, the initial results were counter-productive, for the result of smashing two protons together is usually that a complicated spray of many particles comes out. The ejecta can include protons, antiprotons, and a whole zoo of other particles, most of which decay in such a tiny fraction of a second that even moving at close to the speed of light they travel at best a few centimeters before breaking apart.

The experimenters who made these discoveries in the 1960s did not, of course, bemoan this lack of simplicity. They reveled in it. One thing, at least, was becoming very clear: that the strong "force" was not merely a force in the traditional sense but rather a rich interaction capable of producing new particles in variety and abundance. For more than a decade, new particles were discovered regularly, until they numbered over a hundred. Discovery and classification must precede interpretation and synthesis; things got much more complex before they became simple again.

Over the years, the results of observing many reactions and finding hidden patterns in the combinations of emerging particles were tabulated in a little book full of numbers, all in very small print. The book was (and is) known as the Rosenfeld Tables, after a physicist who devoted much of his career to systematizing the continuing flow of experimental information. Physicists would carry around their Rosenfeld Tables, as others would a Bible or the sayings of Chairman Mao, together with a little magnifying glass to make them readable. The tables grew to include entries for hundreds of different particles, most of them exceedingly short-lived. It was clear that only a major theoretical breakthrough could organize this overwhelming revelation in a meaningful way.

Similar problems have recurred throughout the history of science. There was a problem, dating from the dawn of science, to organize the vast realm of chemical substances and reactions. This realm was successfully organized, in the eighteenth and nineteenth centuries, by the doctrine of elements and atoms. All chemical substances were recognized to be definite combinations of ninety-two (essentially) permanent elements. At a microscopic level, we have ninety-two different kinds of atoms forming various stable associations. Later,

April 1986

Particle Data Group

In addition to the entries in the Meson Summary Table, the Meson Full Listings contain all substantial claims for meson resonances. See Contents of the Meson Full Listings at end of this Summary Table.

Quantities in italics are new or have changed by more than one (old) standard deviation since April 1984.

Particle	$I^G(J^{PC})$ [a] ___ estab.	Mass M (MeV)	Full width Γ (MeV)	Mode	Partial decay modes Fraction(%) [Upper limits (%) are 90% CL]	p [b] (MeV/c)
				NONFLAVORED MESONS		
$\pi^\pm$ π^0	$1^-(0^{-+})$	139.57 134.96	0.0 7.57 ± 0.32 eV	See Stable Particle Summary Table		
η	$0^+(0^{-+})$	548.8 ± 0.6	1.05 ± 0.15 keV	Neutral Charged	70.9 29.1	See Stable Particle Summary Table
$\rho(770)$ $\Gamma_{ee} = (6.9 \pm 0.3)$ keV M and Γ from neutral mode.	$1^+(1^{--})$	770 $\pm 3^g$	153 ± 2 MeV	$\pi\pi$ $\pi\gamma$ $\mu^+\mu^-$ e^+e^- $\eta\gamma$ For upper limits, see footnote e	≈ 100 0.046 ± 0.005 0.0067 ± 0.0012^d 0.0045 ± 0.0002^d seen	358 372 370 384 189
$\omega(783)$ $\Gamma_{ee} = (0.66 \pm 0.04)$ keV S=1.2*	$0^-(1^{--})$	782.6 ± 0.2 S=1.1*	9.8 ± 0.3	$\pi^+\pi^-\pi^0$ $\pi^0\gamma$ $\pi^+\pi^-$ $\pi^0\mu^+\mu^-$ e^+e^- $\eta\gamma$ For upper limits, see footnote f	89.6 ± 0.5 8.7 ± 0.5 1.7 ± 0.2 0.010 ± 0.002 0.0067 ± 0.0004 S=1.2* seen	327 380 366 349 391 199
$\eta'(958)$	$0^+(0^{-+})$	957.57 ± 0.25	0.24 ± 0.03	$\eta\pi\pi$ $\rho^0\gamma$ $\omega\gamma$ $\gamma\gamma$ $3\pi^0$ $\mu^+\mu^-\gamma$ For upper limits, see footnote g	65.2 ± 1.6 30.0 ± 1.6 2.7 ± 0.5 1.9 ± 0.2 0.17 ± 0.04 0.009 ± 0.002	231 170 159 479 430 467
$f_0(975)$ was S(975)	$0^+(0^{++})$	975^c ± 4 S=1.4*	33^c ± 6	$\pi\pi$ $K\bar{K}$	78 ± 3 22 ± 3	467
$a_0(980)$ was $\delta(980)$	$1^-(0^{++})$	983^h ± 2	54^h ± 7	$\eta\pi$ $K\bar{K}$	seen seen	320

A small slice of the Rosenfeld table.

in the early twentieth century, the big problem was to bring order into the realm of atomic nuclei and their reactions. This realm was organized by its own "doctrine of elements," similar in spirit to, but much simpler than, its chemical ancestor. Whereas chemistry required ninety-two building blocks, all nuclei could be built of only two—protons and neutrons.

The next level in deconstructing matter, the problem of organizing the Rosenfeld Tables, was solved along similar lines when, in 1964, Murray Gell-Mann and George Zweig independently proposed a new kind of particles called quarks as the next level of building blocks. Actually, Zweig called them aces, but Gell-Mann's name stuck. The plethora of particles appearing in the Rosenfeld Tables of 1964 were understood as various associations of just three "elements"—three distinct kinds of quarks. These three kinds are denoted by the three letters *u*, *d*, *s*, which stand for up, down, and strange.

The ideas of Gell-Mann and Zweig combined inspiration with audacity. The inspiration was to perceive the patterns concealed beneath the rather chaotic appearance of the Rosenfeld Tables, and the way these patterns could be derived from some simple rules. We will not try to do justice to their inspired guesswork, since to do so would require us to take a long detour into technical detail. We can appreciate some of the <u>audacity</u> involved, though, just by stating the rules Gell-Mann and Zweig proposed. These rules appeared at first—and for almost ten years thereafter—both incredible and bizarre.

First, the incredible part. Quark chemistry contains an interesting new twist that sets it apart from atomic and nuclear chemistry. Atoms, protons, and neutrons are tangible particles you can isolate and study by themselves. Quarks are not. The physical particles, the ones that appear in laboratories and in the Rosenfeld Tables, are all assemblages of several quarks. Quarks join to form strongly interacting particles according to most peculiar *mating rules:* a group may consist either of a quark and an antiquark, or three quarks, or three antiquarks. Other combinations do not occur. As a particular case of this rule, <u>quarks themselves cannot be isolated.</u>

But why not? Once it became clear that the ideas of Gell-Mann and Zweig really "worked"—that they were indispensable to understanding the Rosenfeld Tables—explaining why quarks can't be pulled out of matter became one of the main problems of theoretical physics. Almost ten years passed before the outline of its solution emerged. We'll have much more to say about it as the present theme unfolds.

Now for the bizarre part. The electric charge of every kind of particle ever isolated comes in precise, digital units. Taking the charge on a proton as the standard and calibrating it as $+1$, one finds that every other kind of particle in the Rosenfeld Tables has charge in the same units: some $+2$, some -1, and so on, but always in units showing that the charge could be built up of, or balanced by, an exact number of proton charges. Even the lighter particles, like electrons, bear charges in the same exact units. Protons and electrons differ in almost every other respect—their strong and weak interactions are completely different, their masses differ by a factor of two thousand—but no one has ever been able to measure the slightest mismatch in the cancellation of their electric charges. So most people, if asked to hazard a guess about particles yet to be discovered, would have predicted that their electric charge would consist of some whole number of proton-sized units.

The lightest kind of quarks are the u, or up, quarks. Their electric charge, as predicted by Gell-Mann: 2/3 of a proton unit. Just slightly heavier are the d, or down, quarks, with charge $-1/3$. In fact, all the quarks are predicted to have fractional charges.

If u quarks carry electric charge $+2/3$ and d quarks carry electric charge $-1/3$, then it is easy to see that neutrons

$$\mathbf{n = (udd)}$$

and protons

$$\mathbf{p = (uud)}$$

come out with the right charges: 0 and 1, respectively. It requires a little algebra to demonstrate that, conversely, the neutron and proton charges will come out right *only* if the quarks have these fractional values. But the answer is unavoidable: whole-number charges for the ordinary particles result from, and in fact require, fractional charges for the quarks.

Even if we accept the mathematical requirement that quarks carry bizarre fractional charges, we can hardly stop from once again asking why. Why should the charges of quarks come in such peculiar bits? In this case, too, even the beginnings of an answer did not emerge for almost ten years, and the story is still far from clear. But we're getting ahead of ourselves; we'll take up these matters in chapter 25.

Although up quarks and down quarks suffice to build all ordinary atomic nuclei, other, heavier kinds of quarks show up in unstable particles fleetingly produced by energetic cosmic rays or at accelerators. Particles containing strange quarks have been known since the early 1950s. Other, still heavier kinds of quarks have been found recently, sitting inside even more unstable particles—the c, or *charmed,* quark and the b, or *bottom,* quark. There are compelling reasons to believe that at least one more, the heaviest yet, awaits discovery. It has already been named the t, or *top,* quark. (The names of these things are completely arbitrary, like the names chemists gave chemical elements.) In chapter 30, you'll find a little table giving information about these quarks.

Why are there six kinds of quarks? Are there more? Nobody knows. It is part of a big, unsolved mystery, the mystery of particle families, to which we'll return in chapter 30.

As we mentioned, a detailed description of the evidence that quarks indeed exist would take us far afield. It is like a murder case built on lots of circumstantial evidence—convincing in the end, but laborious to explain. There are hundreds of details in the tables, which quarks organize into meaningful patterns. We will be content to show you just one small glimpse of this exotic realm. Let's see what quarks add to our picture of neutron decays, as discussed in chapter 8.

How does our old friend the decay of the neutron look at the level of quarks? Clearly, underlying

$$n \rightarrow p + e + \overline{\nu}$$

or, in quark language,

$$(udd) \rightarrow (uud) + e + \overline{\nu}$$

there is the basic process

$$d \rightarrow u + e + \overline{\nu}$$

(Figure 19.1 gives this in pictorial form.) Neutron decay is d quark decay, with two extra quarks as spectators.

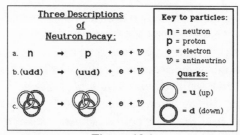

Figure 19.1

So far, we have just introduced another language for describing the same physics. This new description is useful because the same quark process shows up in other places. For instance, according to the peculiar mating rules of quark chemistry, particles should also be formed from the union of quarks and antiquarks—for example, a union of down with antiup, or of up with antiup. Sure enough, particles with the right properties do exist; they are called the pi mesons, or simply pions π^- and π^0. The quark decay above can also be applied when the d quark sits inside a pion. So we predict that the decay

$$\pi^- \to \pi^0 + e + ש$$

must occur, as is shown in figure 19.2.

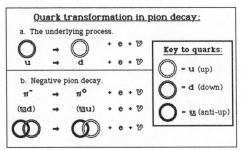

Figure 19.2

The u quark and the ש quark (anti-u) can also annihilate one another, instead of making a π^0. So there is another possibility for π^- decay:

$$\pi^- \to e + ש$$

The expected reactions are indeed found, with the predicted rates. These triumphs are but two of many, many successful predictions that follow from the assumption that the particles appearing in the Rosenfeld Tables are assembled, according to a chemistry following peculiar mating rules, from quarks.

By clarifying the meaning of the Rosenfeld Tables, by translating their jumble of code into a sensible message, quarks allow us to identify the real riddles these tables pose: How do quarks interact with each other? Why do they follow such peculiar mating rules? Why, in particular, are quarks so morbidly shy that they refuse to appear without escorts? The problem of understanding the strong force holding atomic nuclei together is pushed one level down.

20

Colour

The discovery that the elusive quarks come in three distinct colours was surely one of the most remarkable triumphs of the identity-force ideas discussed in chapter 16. For quarks, it was the absence of expected identity forces that required explanation.

Any kind of elementary particle can be defined by measuring just a few of its properties. In the case of electrons, for example, the direction of their spin establishes the only distinctions. Two electrons spinning in different directions experience no identity force; two electrons with the same spin recognize one another as repulsive fermions.

Two quarks can differ not only in spin but also in mass and electric charge. At first, these properties were believed to form a complete set, describing all possible distinctions. Troubles arose, however, because the identity forces expected between quarks identical in every known respect—having the same mass, the same charge, and the same spin—just weren't operating. Quarks were not behaving as proper fermions should.

Some people thought that this meant identity forces did not exist in the subnuclear world. Others, perhaps the majority, concluded that quarks were a mathematical fiction. Quarks seemed incredible and bizarre before, and lack of identity forces between them was the last straw.

But the radical conservatives, who wished to hold on both to identity forces and to quarks, were forced to assume that quarks must have another, previously unknown property. If this is true, the problem with identity forces evaporates. Two quarks with the same mass, charge, and spin could still differ in this extra "hidden variable." Once we are onto this difference between them, we no longer expect them to be driven apart by identity forces.

The extra label that adheres to quarks got called *color*. The name is unfortunate, because quark "color" has nothing to do with color in the ordinary sense. That's why in this book we distinguish it by using the British spelling, *colour*.

An important clue to the number of colours required was found in the Rosenfeld Tables, in a grouping of three up quarks so short-lived it is referred to as a *resonance* rather than a particle. The Δ resonance, or the (**uuu**), shows no repulsive identity forces between its components even when all three up quarks (same mass, same charge) are spinning in the same direction. At least three different colours must be invoked to explain the Δ resonance. In fact, the number of colours required to understand the strong interaction turns out to be exactly three.

The three quark colours are usually called red, white, and blue (although Feynman has suggested lavender, beige, and chartreuse). The colour triplication comes on top of the previous distinctions among quarks; so there are red u quarks, white u quarks, blue u quarks, red d quarks, white d quarks, blue d quarks, and so forth. We should also mention that antiquarks carry anticolours: anti-u quarks can be antired, antiwhite, or antiblue.

Introducing quark colours simply to avoid identity forces may seem to be a desperate and artificial manuever. And it is, of course—but it also brings a great opportunity. If we are taking the idea of colour seriously, we must suppose that nature has formed three different teams (or, if you prefer, three different sexes) for each kind of quark. But so far the only rules we have learned for the colour game are "Red team, don't be identical to the white or blue teams, blue team, don't be identical. . . ." It seems rather pointless to go to the trouble of forming teams, let alone sexes, for such a dull game. Surely, nature must have something more interesting in mind.

And so we are led to ask whether colour is more than a label. In other words, are there forces and interactions that respond directly to quark colours?

A happy question, this, for it conjures up a possible explanation of the peculiar mating rules of quarks. Single quarks are coloured, and single quarks never exist in isolation. Could their colour be to blame? It could, if colour itself brought powerful forces into play. The colour of an isolated quark would then greatly disturb its surroundings until it was somehow balanced or compensated. If the forces called into play are large enough, there will be no peace in the world until the compensation occurs.

The mating rule that permits the union of a quark with an antiquark is easily explained this way. A red quark and an antired antiquark close together would compensate and cancel each other's colours, leaving the world outside happily undisturbed. So such a pair could form a stable unit.

Similarly, if there are exactly three colours, it seems reasonable that three quarks—one of each different colour—can form a balanced set. This would be another possible stable match. In these ideas, we begin to sense a link between the peculiar mating rules of quarks and their colours.

In fact, physicists have heard this tune before. The idea that unbalanced colour brings forth enormous forces recalls a classic old chapter in physics. Imbalance of electric charge brings impressive forces into play. It is the engine of lightning, as Benjamin Franklin revealed with his death-defying kite experiment. As a direct consequence of its power, paradoxically, electric charge tends to hide itself. The powerful forces that charge imbalance calls into play ensure that charged particles will go to great lengths to bind into neutral atoms.

In this way, opposite electric charges work to cancel one another. It is only Heisenberg's uncertainty principle that prevents the electrons from sitting right on top of the nucleus and canceling its charge altogether. Because the charges cancel overall, the atom as a whole does not exert strong electric forces on the outside world. Only when a charged particle penetrates the interior of the atom does it find imperfect cancellation and feel powerful forces. The very power of the electric force makes it self-effacing; it normally operates only in the subatomic world. In that domain, what remain after the cancellation proceeds as far as possible are complicated short-range forces between atoms. These residual forces are responsible for all the complex processes of chemistry.

By changing a few words, we can make this story of the electrical origin of chemistry begin to sound like the beginning of a theory of the strong interaction. Let's make a little dictionary to suggest how the analogy should go:

> **"A Little Dictionary"**
>
> electric charge ↔ colour
>
> nucleus and electron ↔ quark and antiquark
>
> atom ↔ entry in Rosenfeld Tables
>
> chemical force ↔ strong force
> holding   holding
> molecules together   nuclei together

The last entry is most important. Remember that the intimidating thing about the strong force, which made it very difficult to approach, is that it looks so complicated "from the outside in." The lesson to be learned from chemistry, whose inexhaustible richness "from the outside in" arises from utterly simple laws of electricity at the core, is that the complicated strong force between protons and neutrons may be masking simple behavior of their parts.

To put it another way, the traditional idea that protons and neutrons are elementary particles, with complicated forces acting between them, begins to seem completely upside down. It is rather the protons and neutrons that are complicated and the underlying forces that are simple.

21

Gluons

Science seldom advances for long according to the Baconian model, by the straightforward accumulation of facts and classification of structures. As facts are discovered and structures identified, we cannot help playing with them, comparing them to what we already know, guessing what's going to come next.

In this spirit, let's continue to play with the colours of quarks. In the last chapter, we began an analogy between colour and electric charge. Let's see how far we can push it.

What are the essential characteristics of electric charge? First, that it is conserved—we might call this the bookkeeping aspect. In any reaction, the sum of the electric charges of the starting ingredients is the same as that for the final products.

Is colour conserved? Or rather, since there are three types of colours, is each type conserved separately? It would be difficult to justify the importance of colour, or to understand the peculiar mating rules of quarks, if colour had no strict meaning. If the strength of colour charges fluctuated in time, the attractive idea that quark mating rules are simply rules forbidding colour imbalance would lose their solidity. So we have every incentive to try out the simplest and most radical idea, that colour is conserved. (So far, experiments bear this out.)

The second aspect of electric charge is that it governs the electromagnetic interactions of particles. Coulomb's law of electric forces, for example, says that the electric force between two bodies is proportional to the product of their electric charges. This takes us beyond bookkeeping: charge controls dynamics.

Our little theory of the mating rules has already given us hints that colour, like electric charge, is more than a bookkeeping device—that colour governs interactions. Indeed, the basic idea in the theory is that unbalanced colour brings powerful forces into play. Let's try to figure out, by a comparison with electricity, the nature of these forces.

A first guess might be that the strong interaction is just like the electromagnetic interaction, only in triplicate. Each colour would then be like a separate version of electric charge. Each would have its own "electrostatic" force and its own "photon." One "photon" would respond to the total accumulation of red charge in the same way that an ordinary photon responds to an accumulation of electric charge, another to white charge, and another to blue charge.

This first guess is inadequate, however. Recall that a proton, for example, is supposed to be made out of three quarks representing one each of the three colours—one red, one white, one blue. If the strong interaction were merely electromagnetism in triplicate, all three of the colour "photons" would be excited by the proton. The different colours would be not complementary but additive. The forces due to the different colours would not cancel; rather, they would reinforce one another. Since our explanation of the mating rules is that unbalanced colour forces cannot occur, we find ourselves in grave danger of explaining away single protons as well as single quarks.

Similar difficulties arise if instead of three "photons" we imagine a single "superphoton" that responds to the numerical sum of all the colour charges.

A more imaginative generalization is needed. What can you do with three types of colour charge that you could not do with one? Why, you can change one type into another! Only by exploiting this possibility can we forge essential links between the three different colour charges.

We are therefore led to take what will prove to be a momentous step. We must consider new forms of "light," new "photons," which not only respond to their sources but also can transform them.

The colour analogue of the photons will turn out to be not a single particle or a triad but a set of eight. We call it the *gluon octet*, because its members are the glue that hold the proton, and ultimately all atomic nuclei, together.

How can a gluon change the colour of a quark, when each colour charge (we have argued) must be conserved? These two ideas can be reconciled only if gluons themselves are coloured. A gluon that changes a red quark into a blue one must itself bring to the reaction a red charge −1 and a blue charge +1. In this way, the colour charges before the reaction balance precisely the colour charges afterward:

Particles: red quark +	(red+blue) gluon →	blue quark
Red charge: +1	−1	0
Blue charge: 0	+1	+1

To mediate all possible changes of one colour into another, we need six different colour-changing gluons:

Six transforming gluons:		
(red+blue)	(red+white)	(white+blue)
(blue+red)	(white+red)	(blue+white)

In addition, there could be three distinct photonlike gluons that couple to but do not change the three colour charges. In all, it appears we have nine different gluons to consider. One of these turns out, for arcane reasons we'll discuss below, to be unnecessary. Because it is vain to do with more what can be done with less, the possible nine gluons get pared down to an indivisible octet, and one oddball, which does not play any role in the strong interaction.

The couplings of colour gluons are more easily pictured than described in words. In the Feynman diagrams shown earlier, electrons (solid lines) are coupled to one another by wavy lines that stand for photons. A slightly different representation of quarks and gluons is very handy. Solid lines stand for quarks and wavy ones for gluons—but with an important difference. A parallel pair of wavy lines is used to represent a single gluon. These two lines make it easy to follow the flow of colour charge toward and away from the places where gluons interact.

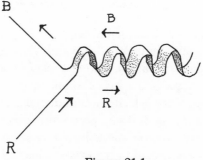

Figure 21.1

Figure 21.1 is a Feynman diagram of the reaction described above, the transformation of a red quark to a blue one by the appropriate colour-changing gluon. The process begins at the lower left, with a red quark moving through both time (vertical) and space (horizontal). Then the quark absorbs a red → blue gluon, which might be said to carry red charge away from the quark and blue charge into it. This interaction changes both the quark's colour and its direction of motion.

The behavior of a gluon that responds to colour, but does not change it, is shown in figure 21.2, as a "plain blue" gluon is absorbed by a moving blue quark. The quark's direction is changed by its collision with the gluon, but its colour is not—the gluon carries blue colour in as well as out. The obvious similarities between figure 21.1 and figure 21.2 reflect deeper mathematical similarities between "responding" and "transforming" gluons.

The fact that gluons themselves are coloured, that they carry colour charges as well as respond to them, means that gluons couple

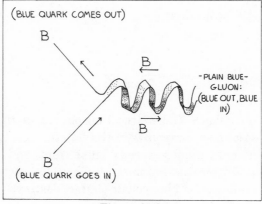

Figure 21.2

to each other. This is a striking and crucial difference between colour gluons and the ordinary photon, of which they are in many other ways a routine generalization. The ordinary photon does not itself carry electric charge; and since photons respond only to electric charge, the consequence is that photons do not directly interact with one another. One photon blithely passes through another; light is to an excellent approximation not deflected by other light. Sorry, *Star Wars* fans, but laser sword fights are bootless affairs.

For gluons, it is a very different story, whose intricate plot is only hinted at by the drawings in figure 21.3. Because gluons themselves carry colour charge, they interact with other gluons, both colour responding and colour changing. The various gluons are connected to one another by a web of different couplings, with dramatic and surprising consequences, as will appear in the next chapter.

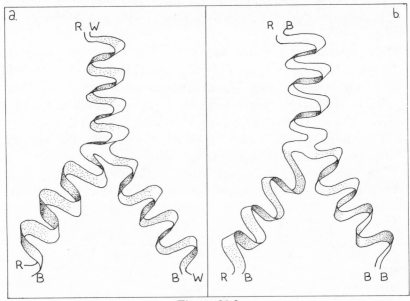

Figure 21.3

The collision in figure 21.3 *(a)* (recalling once again the convention that time passes as we move upward through the figure) begins with two colour-changing gluons—a red → blue from the lower left and a blue → white from the lower right. Their interaction, and the resulting cancellation of blue colour charge, destroys them both to produce a red → white gluon.

In figure 21.3 *(b)*, a red → blue gluon from the lower left collides with a "plain blue" colour-responding gluon. The interaction can be detected only by noticing that the red → blue gluon that emerges has had the direction of its motion altered.

One combination is left out of this scheme. An equal mixture of all three responding gluons responds to the total of all three colour changes—in other words, this mixture makes the "superphoton" we mentioned earlier. Like Superman, the superphoton is an orphan. It does not respond to any of the colour-changing gluons, because the total of all the colour charges for any of these vanishes. The red → blue gluon, for instance, has red charge $+1$ and blue charge -1, so the total of all colour charge is 0. The responding gluons have net charge 0 too. So the superphoton doesn't glue to any of the other gluons.

After removing this mixture from the gluon world, we are left with eight gluons that couple to one another in an indivisible pattern. They form a symmetric octet of gluons, each equivalent to the other in much the same sense that the different sides of a perfect octahedron are equivalent to one another.

Now we can understand why the simultaneous presence of all three colours of quark in a proton is dynamically equivalent to no colour at all. The simultaneous presence of all three colours leads to cancellations between different colours and hence to vanishing couplings to every member of the gluon octet. Since colour forces are due to virtual gluon exchange, no coupling to gluons means no colour forces. So we have been led, in trying to explain the quark mating rules, to guess that a gluon octet exists and mediates the strong interaction in the same sense that the photon mediates the electromagnetic interaction.

With these ideas, the quark mating rules have at last received a deep physical interpretation: No particle that interacts with the gluon octet can exist in isolation. Coupling to the octet unleashes potent forces that are not quenched until, after rearranging particles or if necessary creating new ones, their source is neutralized.

The original example, from which we began, is that single quarks cannot exist freely; their colour must be compensated by other nearby particles. Another one is that gluons themselves, since they interact with other gluons, do not exist in isolation.

Our gluon octet by its very power is highly efficient at hiding anything it can interact with, including notably its own members. How, then, can we tell if it is an elegant and logical fantasy or a physical reality? It is deep within the proton, where quantum uncertainty in position prevents perfect cancellation of colour, that traces of the octet must be sought. To this quest, we now turn.

How Asymptotic Freedom Discovered Me

I (Frank) was a twenty-one-year old graduate student at Princeton—
PRINCETON, where the ghost of Einstein roamed the halls and it
seemed that every second office was occupied by one of my heroes.
I was entranced with the idea of symmetry in physical laws, with the
almost magic power of mathematical group theory in quantum mechanics—and there was Wigner, who started it all! I had just learned
about the Goldberger-Treiman relation, which made a startling connection between the weak and strong interactions—and there, in
adjoining offices, were Goldberger and Treiman! I had reached the
end of the rainbow. And I was floundering.

The kind of problem I was having can be gathered from one of my
typical conversations with a Distinguished Professor:

Distinguished Professor—Here's your desk. You can get pens
and paper down the hall. The library is downstairs. Do something.

Frank W.—What am I supposed to do?

D.P.—Something original and important.

F.W.—But what?

D.P.—Surprise me.

Needless to say, this conversation never really happened. If I had
worked up the courage to ask some of my superheroes for advice
about what to work on (I know now), they would have been delighted
to help. But that was my state of mind at the time. I didn't think I
should talk to them until I had something important to say. So for
almost two years I had been stuck, intimidated, paralyzed.

Great mathematicians can make up problems from pure imagination, but it was rapidly becoming apparent to me that I was not a great mathematician. I needed some problem that would seize me and demand to be solved. I wanted desperately to put my wits—honed on years of mazes, riddles, crossword puzzles, and multiple-choice tests—to work.

At last it dawned on me that what I needed was to get inspiration from the natural world. I thought to myself, "In physics, you don't have to go around making trouble for yourself—nature does it for you."

So instead of continuing to study pure mathematics, I went into the very advanced physics courses, looking for trouble. I was delighted to discover that the most advanced one of all, relativistic quantum field theory, was being taught by David Gross. Someone I had never heard of! That was promising; at least I could walk into the classroom without fear. When I did, I got more delightful surprises: he was young, and he was nervous! Then he started lecturing, and within five minutes two more things became obvious to me. This Gross fellow was very smart, and he cared passionately about the stuff he was talking about. Clearly, this was the guy I was looking for.

I went to all the lectures for this course, did the homework very carefully, and read everything I could find about quantum field theory. There was a huge, indigestible literature, and I didn't understand everything, not by a long shot. But I didn't want to come on like a total idiot. So I tried to sort things out as best I could for a while, made a few minor corrections to little slips in the lectures, and did (if I may say so) a beautiful job on the homework. After a few weeks, I got up enough courage to talk to David; it went something like this:

F.W.—Hi, I've never introduced myself. I'm Frank Wilczek.

D.G.—I know. You're the guy who's doing a good job on the homework.

F.W.—(heart thumping) I've been reading things, and I've got lots of questions.

D.G.—Come on up to my office.

So I went up and opened my notebook and went through the questions. It was wonderful. He was extremely enthusiastic about explaining things. Some of my questions were (in retrospect) vague,

confused, or silly. But he didn't tell me that—instead, he sort of subtly changed the questions into more sensible ones and answered those.

It soon became a regular thing that I went up to his office and talked with him about the things I was reading and got his opinions. These conversations helped me with an enormous problem, of which students or people who look only at standard textbooks and sanitized histories of science are usually unaware. In the textbooks, only the successful ideas are recorded. You get the impression that the history of science is a totally progressive, orderly, logical development of ideas. But when you read research at the frontiers of knowledge, you quickly realize it's not that way at all. Different people, all respected experts, are trying completely different and mutually contradictory approaches. The good ideas, the ones that eventually find their way into the textbooks, are out there, all right—but they are out on the shelf surrounded by all kinds of junk, which is packaged by its authors as attractively as possible and, of course, is never clearly labeled "junk." It's not that people are out to fool you. They may be just confused or mistaken. Or, more commonly, they are doing work that is correct, in the sense of not containing errors, but not useful for progress. The work is too speculative or too safe; it loses contact with reality or remains narrowly earthbound.

To be able to sense what's genuinely promising takes a lot of experience and intuition, and nobody is perfect at it. But the ability to sniff out the promising directions is at least as important as raw intelligence for doing research. David would often say that some idea "smells right" or that some approach "just doesn't smell right." At first I thought this was ridiculous, but after spending some time in his office, I began smelling things too.

Before long, our noses were leading us along a definite trail.

Some experiments at the Stanford Linear Accelerator (SLAC) had seemed to indicate that the forces between quarks turn off when the quarks get close together. If this was the right interpretation of their results, it was clearly a tremendous clue we would need to take into account in understanding the strong interaction. On the other hand, the idea that interactions "turn off" at short distances was very hard to reconcile with other theoretical ideas. We certainly hadn't had any prior experience with forces behaving that way. All the forces we knew about, like gravity and electricity, get bigger as you bring two bodies close and "turn off" only when you separate them. In fact,

the weird, topsy-turvy behavior implied by the SLAC experiments seemed likely to contradict fundamental principles of relativity and quantum mechanics. In analyzing what happens at short distances, you have to worry about the very wild fluctuations we talked about before, in connection with antiparticles. To put it another way, a dense cloud of virtual particle-antiparticle pairs immediately surrounds any particle. The stuff in this cloud, which gets denser and denser as you get close to the center, interacts too. In fact, in every theory that anybody had ever calculated, the effect of the interacting cloud was that forces got larger and larger—out of control—at short distances.

By the time I came on the scene, this effect had been calculated for a wide class of possible kinds of interactions, always with the same result: the forces got larger at short distances. It was beginning to seem that a serious crisis was brewing. Either the SLAC experiments (which seemed very solid) or sacred principles of relativity and quantum mechanics were at risk, because it seemed very hard to have both.

There was a loophole that hadn't been closed, however. A particularly tricky class of theories, with the horrible name *nonabelian gauge theories,* had not been examined. There were two major reasons for this neglect.

First of all, they had only very recently become respectable. You see, the formulation of these theories requires that you consider particles created with negative probability in some of the intermediate stages of the calculations. Real-life probabilities don't get any lower than zero. Clearly, if these negative-probability particles escaped from your scratch paper into the real world, the theories were nonsense. Probabilities for real events can be small or zero, but not negative. For many years, no one knew whether the negative-probability particles were a mathematical crutch or a real disease in the theory. In other words, it wasn't known whether if you started with only positive-probability particles and let them collide, only positive-probability particles would come out. Finally, Gerhard t'Hooft, a Dutch graduate student, proved this in 1970 (the theories had been around since 1938 and popular since 1954).

Second, there was the sheer technical difficulty of working with these theories. They contain lots of particles, among them the bizarre negative-probability ones, coupled to one another in intricate ways. This makes it tedious to calculate even the simplest physical

processes in these theories and, of course, means it's hard to avoid mistakes. (Or, even if you have avoided them, to be sure you have.) There is also a more complex problem that goes under the rubric "maintaining gauge invariance." The essence of this problem is that the most convenient formulation of the theory contains a lot of stuff (like the negative-probability particles) with no direct physical meaning. And this means that it is easy to start calculating pure nonsense without being aware of it.

David and I set about trying to close the loophole. We were going to calculate whether the interactions turned off at short distances in nonabelian gauge theories.

For David, this was the last part of his program to destroy quantum field theory. His eyes seemed to light up as he relished the idea with maniacal glee: "I'm going to prove that it's impossible, absolutely impossible, to have scaling in quantum field theory." (*Scaling* is a shorthand description of the results that the people at SLAC had observed.)

For me, it was the perfect thesis problem. The necessary calculations were technically challenging, yet guaranteed to lead to definite results. I would learn a lot and have something definite to show for my efforts, whichever way it went.

The calculations I needed to do would be lengthy, but basically straightforward, with modern techniques. At the time, however, the theoretical tools available were much more primitive. First, it was necessary to convert the seemingly simple word problem "What is the strength of the force at short distances?" into unambiguous mathematical formulas, picking a path through the dense forest of unphysical clutter that appears in gauge theories. And then, of course, it was necessary to evaluate the resulting formulas, which required a lot of care and a few new tricks. All this took quite a while, and I filled a thick notebook reproducing the old known results for nongauge theories and experimenting with different approaches to the problem at hand—gauge theories. This activity went on through the fall of 1972, and I was consulting with and getting plenty of help from David all through this period.

By winter, I was starting to get definite numbers, but I could see they were rubbish. A feeling of déjà vu swept over me, as I recalled one of my earliest childhood memories. It was when I was four or five years old, and just starting to learn about numbers and how to make change. I had a brilliant inspiration (I thought) that would make my family rich. The idea was to start with pennies, exchange them for nickels, exchange the nickels for dimes, then go to dollars, to quarters, back to dimes, and so forth, in various complicated ways, and eventually change it all back to pennies. You see, if you wound up with more pennies than you started with, you could do it again and again, and make yourself rich. The true beauty of the idea is that the contrary result, fewer pennies at the end, is equally good—just do it backward! Now, I wasn't very good at the calculations, and I came up with quite a few schemes that seemed to work. I proudly showed my little notebook full of numbers to my father. He was amused but rather skeptical. He went through the calculations with me and always managed to find some mistakes. When we fixed the mistakes, it always came out the same number of pennies. Then he explained to me that it really had to be that way; after thinking about it, I became reconciled to this disappointing truth.

The problems I was having with my gauge theory calculations were really quite similar. I would calculate how the forces were changing for spin 0 particles and get one answer, for spin 1/2 another, for gluons (spin 1) another, for ghosts (the negative-probability guys) yet another. But that couldn't be—the forces should depend only on colour charge, period, no matter what kind of particle carried the charge. The answers should always be the same. It was just as in the old days, when I got different amounts of money by calculating in different denominations. There had to be a mistake—or, rather, several mistakes—in the calculations.

So I began the tedious chore of debugging my calculations. A major breakthrough came when I realized that one of the input numbers I had taken from a standard reference had a sign error. I had wasted many hours in consternation and confusion, simply because I was too trusting (and a little lazy) and hadn't taken five minutes to check this thing. Once this glitch was corrected, everything was close to agreeing. I could begin to see light at the end of the tunnel, but there were still some inconsistencies to sort out.

During this period, a couple of scary things happened. Just after the winter break at Princeton, in late January, a preprint arrived from the famous German physicist Kurt Symanzik. It was the written version of a talk he had given at Marseilles in the summer of 1972. In this little paper, he discussed how nice it would be to have a field theory with scaling. I could tell that's what he was talking about, because we were working on closely related ideas, although his formulations were extremely obscure. (Other physicists have told me they found it very hard to make head or tail of Symanzik's papers, and I can understand why. The difficulty people had in understanding Symanzik powerfully affected the history of physics, as we shall see.)

Anyway, in this paper he actually discussed an example of a theory with scaling. His example was seriously flawed, as he recognized, because the theory he was using is unstable. In his example, the attractions between particles were such that you could gain more energy by bringing a lot of them together than it cost you to make them in the first place. So, in this theory, empty space is an explosive mixture. Nevertheless, Symanzik implied, you should ignore this slight problem long enough to notice that the theory had a very good property. He could show that in this theory the interactions between particles turned off as they got close together—just what we wanted, in order to understand the SLAC results. All this was stuff David and I knew, but it was slightly scary to find that other people had been on the same track months before, and now the whole thing was becoming public knowledge. Far worse, as far as I was concerned, was a sentence at the very end of the paper that said something to the effect "it would be very interesting to know how nonabelian gauge theories behaved in this regard." As I read this, I got a sinking feeling in the pit of my stomach. I saw my thesis, and months of work, going down the drain.

The other scary thing was that David told me a student of Sidney Coleman's was doing the same calculations, for other reasons.

(Sidney Coleman was visiting Princeton, on leave from Harvard, while all this was going on. Sidney is a wonderful and remarkable man. He looks like Albert Einstein and talks like Woody Allen. For many years, he gave lectures at Erice, Sicily, every summer. In these lectures, he refined the best ideas from the preceding year in theoretical particle physics into pure, luminous gold. Transcripts of Sidney's Erice lectures formed a central part in the education of a whole generation of physicists, including me. Having Sidney at Princeton during this period was very important to us, both because he kept us honest about gauge invariance and—even more important— because we felt that if Sidney was interested in what we were doing, we must be doing interesting things.)

These two scares made me pick up the pace. An intense period began, during which David and I spent hours a day, every day, straightening out the calculations. Most of the pieces were in place, but it took only a few errors to foul up the whole thing. A bookkeeper who has to add hundreds of numbers can't be content to get 95 percent of the additions right—his books won't balance. We had to get it perfect. After about two weeks of intense checking, we got all the different ways of calculating the colour force—for different kinds of particles, as I mentioned before—to agree.

At this point, we made one last, final, comic mistake. After working so hard to get all the different methods of calculation to agree, we were exhausted and got very casual about checking the sign of any one of them. So for a few days we thought the force got stronger at short distances for nonabelian gauge theories, just as it does for all other kinds of theories. In the process of writing the results up carefully for publication, David found that he kept getting the other sign. Very early one morning, I got a very excited and mysterious phone call at home from him. (Actually, it was around noon, but in those days I worked late into the night and was usually asleep at noon.) From the phone call, I expected him to be delirious when I arrived, but instead he was eerily calm: "We'd better check this. I think we have asymptotically free theories." *Asymptotically free* theories are ones in which the force turns off at short distances, just the thing we knew we needed to understand the SLAC results. Once we were alerted, it was a very easy matter to check the sign, and within half an hour there could be no doubt. We had found asymptotically free theories.

What's more, the theories were almost unique. The result was perfect, a gift from heaven. If you took both the experiments about scaling and the general principles of quantum mechanics and relativity seriously, you were led to a very specific and complete theory of the strong interaction.

The magnitude of this discovery didn't immediately sink in for me. At this stage in my career I didn't yet realize how frustrating theoretical physics usually is and how rarely anything works. There is a bit of physics folk wisdom (I don't know where it originated) to the effect that "if a theory is simple, elegant, and not obviously wrong, then it's probably right." This is an exaggeration, no doubt, but after experiencing at first hand how hard it is to introduce new ideas that are at once simple and consistent with everything that's known into the tight structure of physics, I understand the feeling very well. So at the time I wasn't as convinced we had solved the historic problem of the nature of the forces responsible for the strong interaction—marking the beginning of the end of a fifty-year search to figure out what held atomic nuclei together—as I probably should have been. But no matter; I didn't need to know the theory was right to be happy working on it. I was happy, as ever, to be doing little puzzles—and delighted that other people around Princeton were getting quite excited about our work.

Anyway, at this stage—with some interruption in the summer, when I got married and David went to Fermilab for a while—we began an even more intense period of work, to classify all possible asymptotically free theories and to put them to use.

Our interactions did not simply "turn off" in an abrupt way at short distances; instead, they faded out in a very structured way that could be tested experimentally. We could tell the people at SLAC, and other people doing follow-up experiments at other laboratories, what to expect if they did their experiments more accurately and at higher energies. To do this took quite a bit of calculation, but it was pretty straightforward after the exertions of the preceding few months. We made a game of it, racing to see who could calculate the graphs faster. David was usually faster, but when our answers disagreed mine was usually right. (David may give you a different story, but that's how I remember it.)

At first, and for about a year, the experiments looked very bad for us. People smashing electrons and positrons together at high energies were finding many more interactions than we expected. We predicted that the rate of interactions should go down dramatically with rising energy, but the experimental results seemed to show a constant rate. All this changed virtually overnight, in late 1974, when it was discovered that the unexpectedly large number of interactions were due to the production of a new kind of quark, the charmed quark. (The existence of this new kind of quark had been predicted shortly before its discovery by Sheldon Glashow, John Iliopoulos, and Luciano Maiani. The name was inspired by the traditional use of charms to ward off evil, because the existence of this quark was necessary to ward off some terrible problems in understanding the details of the weak interaction.) Our theory had a lot to say about how quarks interacted, but nothing about how many quarks there were. So the discovery of charm, while it had nothing directly to do with our theory, removed what had seemed to be a serious difficulty.

In the twelve years since, QCD—as the modern asymptotically free theory of the strong interaction is now called—has gone from success to success, and it is now universally recognized to be the theory of the strong interaction. At today's accelerators, quarks and gluons leave their signature in an extremely legible form—you can practically see them in pictures of the stuff emerging from ultra-high-energy collisions, indirectly but quite clearly. They leave narrow trails of particles in their wake, like the plume a jet plane writes across the sky.

I have mixed feelings as I look at such pictures. On the one hand, I think, "Isn't it marvelous. We were right back then. Nature works just the way we thought." And on the other hand, "We really were pretty useless. After all, everything would have become obvious from these pictures, even if we had never calculated anything." Which attitude is right? If science is defined only by its objective results, no doubt the second is. But I think science is more than its results, beautiful and important as these are. It is also the human effort and feeling that goes into them. So I don't feel completely silly in taking personal pleasure from these pictures.

Finally, a few words about "history." The above account is very much a personal perspective, based completely on my recollection of what I went through at the time. Other physicists made important contributions I was unaware of at the time, and they might understandably have very different ideas about which developments were crucial.

I'd like to mention in particular Tony Zee, who published a paper containing all the right physical ideas and outlining the required calculations. He wanted to check nonabelian gauge theories, too, but stumbled amid the technical difficulties.

Gerhard t'Hooft, who (as we mentioned) developed the framework we relied on, did calculations closely related to ours. I later learned that he was actually present when Symanzik gave his Marseilles lecture and that he remarked at the time that the interactions would turn off in nonabelian gauge theories. Apparently there was a failure of communications, though, because t'Hooft didn't publish his results until well after we published ours, and he didn't seem to recognize their crucial physical importance for the theory of the strong interaction.

And the "Sidney's student" scare turned out to be quite justified, because the student—David Politzer, now a professor at Caltech— got results very similar to ours at the same time.

So, clearly, asymptotic freedom was waiting to be discovered, and if we hadn't done it, physics wouldn't have been much delayed. I don't want to be too humble about this—David (the two Davids, actually) and I did some very clever things and worked extremely hard. But I've always felt the discovery was, in a sense, the other way around— and I'm eternally grateful that asymptotic freedom discovered me.

22

Asymptotic Freedom

How often have I said to you that when you have
eliminated the impossible, whatever remains,
however improbable, *must be the truth?*
    —A. C. Doyle, (att. Sherlock Holmes),
                    The Sign of Four (1888)

Nature poses many riddles but contains no contradictions. By solving one of her puzzles, therefore, we are guaranteed to learn something—and the weirder, the more impossible the paradox seems at first, the more mind-expanding will be its ultimate resolution.

Among all the paradoxes nature has dreamed up to delight the theoretical physicists, none are more attractive than her unanticipated simplicities—findings of unity and order where complexity or chaos seemed inevitable. Both theories of relativity sprang from such paradoxes: the special theory, from the paradox that light's speed is independent of the speed of its source or receptor; the general theory, from the paradoxical "coincidence" that all bodies, regardless of their mass, move the same in response to gravitational forces.

The key to unraveling the nature of the strong interaction was a paradox of this kind. The enigmatic behavior in question is this: although the strong interaction generates the most powerful forces we know of, it seems in certain situations to turn itself off—to vanish completely.

This paradox was first made clear by a very convincing and complete set of experiments done at the Stanford Linear Accelerator (SLAC) in the late 1960s. These experiments were set up to look at effects of the strong interaction in high-energy collisions between electrons and protons.

We have mentioned earlier the correlation, when you're trying to look at particles, between high energies and short distances. The idea was that electrons would smash into protons so hard that the quarks themselves—or whatever lurked inside a proton—would be disturbed and lots of gluons—or whatever was holding nuclei together—would come radiating out of the collision. In other words, some of the collision energy would be used up by exciting the quanta of the colour field.

By looking at the outgoing electrons, and seeing how colour effects had reduced their energies and changed their angles, scientists hoped to get information about the colour forces. No simple answers were expected to emerge, but just because the strong interaction is so strong, its effect was expected to be very noticeable.

The corresponding process for electricity—smashing charged particles together and exciting quanta of the electromagnetic field (photons)—was already quite well understood. Taking this effect into account, people calculated how the electrons ought to come out if there were no strong interaction at all. The big surprise was that that's almost exactly how they do come out of the collisions—as if there were no strong interaction at all.

This paradox—that at very short distances nature simplified matters by letting one of her strongest, most complex forces disappear—seems even stranger in contrast to the way electromagnetic forces behave at short distances. When two charged particles are brought closer and closer together, the electric forces between them just get stronger and stronger.

In fact, there's an effect in electromagnetism known as *screening* of charge by the vacuum that makes the paradox presented by the SLAC experiments even sharper.

The basic idea that an electric charge gets screened by its surroundings goes back to the dawn of experimental electricity, to Faraday. Before we look at how this can happen even in a vacuum, it's helpful to start with something a bit more tangible—a positively charged ion surrounded by water molecules, as shown in figure 22.1.

Water molecules get electrically polarized because electrons are attracted away from hydrogen by the oxygen atom. So the oxygen atom in each water molecule ends up with a little bit of extra negative charge hovering around it; the hydrogen atoms have a little bit of a positive charge. When a positive ion is dropped into this dynamic

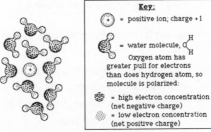

Figure 22.1

situation, the water molecules respond at once. There is, despite the constant, random thermal jitter, a tendency for a water molecule to line itself up so that its negative end (the oxygen atom) points toward a positive ion and its positive ends (the hydrogen atoms) away.

The positive ion, surrounded by negative charges, has its electric effect shielded and partially canceled. We say the ion is *screened.* Only very close to the ion will a particle be able to feel its full electric field. Because of this canceling effect of the medium, particles far from the ion will sense an "effective charge" quite a bit smaller than the bare charge of the ion itself.

We know now something Faraday himself with prophetic genius anticipated: that empty space—what we call the vacuum—is not a passive receptacle devoid of properties but a rich dynamic medium. It is filled, for one thing, with *virtual particle-antiparticle pairs.* These are quantum-mechanical fluctuations, which can be thought of as matched particle-antiparticle pairs that fleetingly come into being, only to annihilate without a trace.

The virtual pairs, though transitory, are not without effect. Consider how they interact with a positively charged ion. While a pair exists, its negatively charged member is attracted toward the ion, and its positively charged member is pushed away. Such pairs, in short, behave just like the polar molecules in a more conventional medium. Although any particular virtual pair quickly disappears, new ones are constantly appearing, so their average density stays constant. Virtual pairs make the vacuum a dynamic medium, with the ability to screen electric charge.

Screening by the vacuum has even been observed experimentally. If an electric charge is partially screened by the vacuum, then at very short distances, where screening is less complete and other particles can more nearly feel the whole "bare" charge, electric forces are correspondingly larger. As one consequence, the force between two charges should rise steeply but predictably as the distance between them gets very small, not only because the charges are getting closer to one another but also because they are penetrating one another's screening clouds. Now, the force law is used to calculate the energies of electron modes and, from these, the wavelength of spectral lines. Both the calculations and the measurements can be done with great precision. When these are compared, it turns out that they agree only if vacuum screening of the electric charges is included. This is impressive evidence that the vacuum—space as empty as it can possibly be—is filled with virtual pairs that screen electric charge.

Back to SLAC. If vacuum screening also occurs for colour charge, what will happen when an electron scatters off a quark? The quark itself (that is, the bare charge of the quark) responds immediately to the impulse supplied by the electron, but the surrounding particle-antiparticle cloud takes a little time to reorganize. (For one thing, the message that the quark has moved cannot be propagated faster than the speed of light; this already implies that the cloud has a finite "reaction time.") So the quark emerges stripped of its usual screening cloud, carrying just its bare colour charge. Since the bare charge is stronger than the screened charge we normally see, we should expect that in high-energy collisions the strong force would look even stronger than usual. So, trying to use the electromagnetic model of screening by the vacuum to predict quark behavior leads only to a deeper contradiction of the SLAC experiments.

Within this debacle, however, we perceive the germ of a possible resolution. What we need is the opposite of screening. We need a theory in which colour charge, instead of being screened, is antiscreened by the vacuum; in other words, a theory in which a small bare colour charge polarizes the surrounding vacuum so as to reinforce rather than shield itself. If this occurs, the effective charge we ordinarily encounter in strong-interaction processes will be larger than the bare charge of the central quark (without its reinforcing cloud). The seeming paradox, that particles that interact strongly when well separated radiate only weakly when struck, will then become comprehensible.

It is not at all clear that there is any sensible theory with antiscreening behavior. In this context, "sensible" theories are those that do not force us to trample upon well-tested principles of physics. We would not like to give up basic principles of quantum mechanics or relativity, for example, to achieve vacuum antiscreening. Before considering such desperate measures, we will do well to try the radically conservative strategy of building upon, rather than dismantling, the existing framework. As Feynman likes to say, we must use our imagination, but imagination in a terrible straitjacket. There is also an enormous practical advantage to working this way: we know, more or less, what the rules are.

It turns out that the rules for formulating theories consistent with general "sacred principles" such as causality, the standard rules of quantum mechanics, and the theory of relativity are very limiting. That is the flip side of the power of these principles. For instance, we have discussed before how electromagnetism, which describes, as Dirac says, "all of chemistry and most of physics," is completely dictated by a simple basic coupling between the photon field and the electron field.

The framework restricts us as we seek to describe the strong interaction; it also supports our efforts. From the general principles of relativity and quantum mechanics, we have learned to describe the world in terms of fields. These general principles are so powerful that to specify a physical theory completely we need only state what fields exist and give a schematic description of their couplings. In the language we used before when discussing fields, we need only know the chemicals in the universal ocean, and the reactions among them.

There are only a few possible forms of basic couplings. Each kind of coupling has definite physical consequences, which can be systematically worked out. In particular, it is possible to calculate whether the sources of the interaction, the fundamental particles, are screened or antiscreened—whether vacuum fluctuations cancel or reinforce their power—case by case. This is a well-defined, though rather formidable, mathematical problem.

A wonderful result emerges from these calculations. It is captured in the mathematical symbol ∃!—a sort of ∃ureka!—meaning the answer exists and is unique. Only one class of theories has antiscreening behavior. And the theories that are picked out are just theories of coloured quarks interacting with gluons, the same theories we guessed before just by classifying particles and playing with analogies. A wonderful harmony between form and function, classification and dynamics, becomes apparent.

Vacuum antiscreening of colour charges is called *asymptotic freedom*. (The word *asymptotic* contains the idea of getting closer and closer to zero without ever actually reaching it.) This name is used because at asymptotically small distances quarks behave as if they were free of the strong force.

Asymptotic freedom or, equivalently, vacuum antiscreening takes some getting used to. It is, of course, diametrically opposite to the screening behavior we are familiar with (and can easily explain) in ordinary electricity. Somehow the physical consequences of having just one colour—that is, a single type of charge, as in electrodynamics—are dramatically different, in this key respect, from what occurs when you have two or more.

We can relate asymptotic freedom to the major qualitative difference between one colour and many. Unlike photons, the colour gluons are themselves charged and linked by an intricate pattern of cross-couplings. In considering how virtual pairs react to a charge source, we must take into account virtual gluon pairs, whose presence (because of the cross-couplings) affects other virtual pairs. In figure 22.1, we sketched a simple qualitative picture of the way a bare electric charge is screened by its surroundings. To construct a similarly simple picture for the mechanism behind antiscreening of colour charge is quite beyond our artistic powers.

Asymptotic freedom leaves us poised at the brink of instability. Antiscreening means that a little charge at the center is reinforced by the vacuum, looking larger from a distance and (iterating) still larger from a larger distance. Where does it all end? The buildup of colour charge can end only if a neutralizing anticharge or complementary colour charges are available to cancel the charge of the source. Asymptotic freedom at very short distances leads to a heavy burden of charge at longer distances. Though it is hard to calculate exactly what happens, it does not seem farfetched to suppose that an unlimited buildup of charge is forbidden and that canceling sources will always be pulled in from nearby, or if necessary conjured up by the creation of particle-antiparticle pairs, to prevent it. With this thought, we arrive once again back at the quark mating rules, now really understanding them for the first time.

We hope this discussion has conveyed the beauty and inner coherence of the idea that the strong interaction is ultimately described by the asymptotically free theory of coloured quarks and a gluon octet. A good scientific theory must, however, do more than give a consistent account of the observations it was specifically designed to explain. Before we can regard the colour theory of the strong interaction as established, we must draw new consequences from it and compare them with experimental realities.

Fortunately, thanks to asymptotic freedom, our fundamental theory of the strong interaction is not only deep but also fruitful. The theory claims that, at short distances, the strong interaction becomes not only weaker but much simpler. This suggests several ways in which the basic ideas may be applied and tested, both qualitatively and quantitatively. Many successful applications have been made, and the theory has never failed a test.

One kind of test involves the corrections to perfect freedom: if a quark is accelerated less than infinitely fast, as, of course, it always is in practice, it carries part of its reinforcing cloud with it. The less extreme the acceleration, the better the cloud will follow the quark.

A slowly accelerating quark, then, carries with it a large amount of colour charge, couples strongly to the gluon field, and has a good chance of radiating. Asymptotic freedom is, as advertised, only asymptotic. At any finite acceleration, there is a finite probability of radiating, a correction to perfectly free motion. These corrections have been calculated, and their predictions for the patterns of scattered electrons have been compared with the findings of many experiments at SLAC and elsewhere. The agreement between prediction and reality is impressive.

Another class of tests and applications concerns *jets*. When quarks or gluons are suddenly accelerated, we have argued, they tend to fly right out of their antiscreening clouds. But eventually the cloud left behind will respond to the departure of its reason for being by disintegrating; conversely, the "vacuum" near the moving quark will organize a new cloud. These extensive rearrangements in the world of virtual particles are accompanied by the creation of many quark-antiquark and gluon-antigluon pairs. A shower of particles—protons, antiprotons, pions, and so forth—along the path of the quark is the observable result. Such a shower of particles all moving in roughly the same direction is, for obvious reasons, called a jet.

Jets are complicated relics of obscure processes. Fortunately, they retain some record of the simple properties of the quark or gluon that triggered them. Quantities conserved by the fundamental interactions—such as energy, momentum, and direction of motion—are, almost by definition, not altered in the process whereby a quark or gluon produces a jet. Each jet is a narrow spray of particles, following the trail blazed by the parent quark or gluon and sharing its energy and momentum. Many detailed properties of the jets, such as their total energies, relative angles, and the number of jets that occur, can be calculated. Once again, the agreement between predictions and experiments has been outstanding.

Collisions at modern high-energy accelerators give rise to beautiful jets. Looking at pictures of them, we come to sense that quarks and gluons, although they cannot be isolated, are no less real than electrons and photons. In a remarkably direct and spectacular way, we can "see" them in the jet trails they leave behind.

In Praise of QCD

Our search for a deep understanding of the force that holds atomic nuclei together, the strong force acting between protons and neutrons, has led to a strange outcome. We started with a tangible force—something that exerted pushes and pulls between tangible particles—and wound up with a theory of transformations among coloured quarks and gluons, particles (the theory tells us) we shall never be able to isolate. In the preceding chapters, we saw how this outcome is forced upon us by experimental realities and theoretical analysis. Let us pause briefly to survey the new territory won, to appreciate the splendor of the quantum theory of colour (quantum chromodynamics, or QCD), and to view it in a wider context.

Perhaps most important, QCD makes the strong interaction begin to look more like the other known interactions. We have already been at pains to emphasize that the basic interactions of QCD are direct and natural generalizations of quantum electrodynamics (QED). For example, colour charge is a generalization of electric charge; the octet of gluons is a generalization of photons; protons, neutrons, and the many other denizens of the Rosenfeld zoo are constructed out of quarks and antiquarks, much as atoms are constructed out of nuclei and electrons.

There is a subtler, but no less important, resemblance between QCD and the weak interaction. The transformations built into the fabric of QCD mean that the weak interaction, which previously seemed to be an isolated example of a transforming principle, is no longer isolated.

These connections give us hints of a possible unified understanding of what previously seemed utterly disparate kinds of interactions. In the following pages, we shall see that impressive progress has been made in building on these hints.

In the 1960s and early 1970s, a large school of thought held that the strong interaction was a world in itself, which would require radically new concepts of the order of quantum mechanics for its understanding. It was even hinted that somehow Eastern mysticism had something to teach us here. Well, Eastern mysticism may or may not have something to teach us, but we certainly don't need it to describe the strong interaction. QCD has been a radically conservative scientific revolution, pushing the basic concepts of quantum mechanics and relativity into new domains and exposing (for instance, in the concept of asymptotic freedom) some of their unanticipated richness, but not overthrowing them.

Asymptotic freedom has an important application to big bang cosmology. In the very early stages of the history of the universe, densities and typical energies of the soup of particles that filled all space were very high. The quarks were close together. Asymptotic freedom tells us that their behavior in these circumstances becomes <u>simpler</u> to describe; closely spaced quarks sense only one another's small bare charge and hence propagate nearly freely. It becomes possible to extrapolate back to times within a millisecond of the big bang, to densities much higher than nuclear and temperatures above 10^{14} degrees Kelvin, with some confidence that we understand the behavior of matter in these extreme conditions.

Apart from these "applications" to other domains of physics, QCD has a remarkable inner perfection. In this respect, it surpasses even QED, its ancestor and model.

QED reduces—that unfortunate word, not intended at all pejoratively here—an enormous wealth of phenomena, "all of chemistry and most of physics," to one basic interaction, the fundamental coupling of a photon to electric charge. The strength of this coupling remains, however, is a pure number, the so-called fine-structure constant, which is a parameter of QED that QED itself is powerless to predict.

QCD reduces, in a similar sense, nuclear physics to a single, basic coupling between gluons and colour charge. It goes QED one better, however. In QCD, it is not necessary to supply a number that specifies the strength of the coupling. Because of the antiscreening property, the colour charge, and hence the coupling strength, of a quark depends critically on the distance within which you measure it, and actually vanishes in the limit of infinitely small distances. We can use any coupling strength we like, with the understanding that it is the coupling strength at a particular distance. Changing the coupling is equivalent to changing the unit of length. Just as nothing fundamental can depend on whether you choose to measure distances in inches or centimeters, these "different" versions of QCD all lead to the same predictions.

While QED is a theory that predicts much in terms of one parameter, QCD predicts much in terms of no parameters at all.

Eighth Theme

Symmetry Lost and Symmetry Found

The basic idea of symmetry is that you can change an object in some way—by rotating it, say, or reflecting it in a mirror—without changing the way it looks or acts. Artists often use symmetry as an organizing aesthetic principle. Sometimes, as in Islamic art or the work of Escher, symmetry as such occupies the center of attention.

Physicists too employ symmetry as an organizing aesthetic principle. It is wonderful to find that fundamental physical laws do in fact exhibit a high degree of symmetry. In other words, we can (in our imaginations) change the world without changing the way it operates.

Recently, symmetry has come to play an ever more central role in physics. In new and unknown domains of the microworld, physicists often try to guess the laws by <u>demanding</u> symmetry (instead of patiently finding the laws and then <u>noticing</u> their symmetry). This bold procedure has proved so fruitful that on those occasions when reality fails to live up to expectations of symmetry, it has become customary to hold on to the symmetry by inventing imaginary worlds. Reality is identified with an imaginary world of greater symmetry, seen, as it were, through distorted lenses. Here we examine the imaginary world, and the lenses.

Prelude Eight

RELATIVE AND ABSOLUTE

"What's in a name?" Nothing, or very little, if the name is merely a symbol of something already known, as Shakespeare has Romeo inform us: "A rose by any other name would smell as sweet." The situation changes, however, if you know the name of something, but little else about it. Then the name looms larger. If roses were called skunkweeds, someone who hadn't smelled one before might have mistaken expectations.

Unfortunately, the theory of relativity has a name that suggests interpretations the theory itself does not support. The theory's impact on the popular and literary imagination has, sadly, often been determined at least as much by the name as by the thing itself. One main effect of the name has been to lend a most undeserved aura of intellectual respectability to such lame expressions as "Well, it's all relative to the way you look at it, after all," used as an excuse to avoid difficult judgments and comparisons. At a slightly higher, or at least more insidious, level, we encounter intellectual movements and ideas that borrow the name: "ethical relativity," the idea that there exist many ethical systems of equal validity (in the extreme version, one for each human being), and "cultural relativity," which is much the same thing, but applied to cultures rather than individuals. We are not going to argue the merits of these views here, but we will try to make it clear that their use of the term *relativity* is a grotesque distortion of its use in physics.

One of the seminal sources of modern relativity theory was a lecture entitled "Space and Time," delivered in 1908 by the distinguished German mathematician Hermann Minkowski. This lecture

was of considerable historical significance. It marked the first exposure of the scientific community at large to Einstein's new ideas.
That these startling ideas were publicly endorsed by an internation-
ally renowned mathematician before a large international confer-
ence (at a time when Einstein himself was still a very young, little-
known figure) was an important step in making them "respectable."

Yet in his famous lecture, Minkowski said that the name "relativity postulate" for the central axiom of the theory seemed to him
"very feeble," and he went on to state, "I prefer to call it the *postulate of the absolute world.*" Max Planck, another early advocate of
relativity theory and the founder of quantum mechanics, expressed
similar sentiments. If you are accustomed to thinking of relative and
absolute as concepts diametrically opposed, this argument about
whether to call a single postulate, on whose content everybody
agrees completely, the relativity postulate or the postulate of the
absolute world may at first seem bizarre or even surreal. But a
deeper consideration of just what the theory actually is reveals that
both names capture important partial aspects of it.

Relative and absolute are here two sides of the same coin. The coin
is *symmetry.* The idea of symmetry has come to occupy a central
place in our description of the physical world; appreciating its dual
manifestations, as relative and absolute, is a good entry into understanding its meaning and use.

The postulate whose name was debated between Einstein and
Minkowski is this: the laws of physics will appear the same to any
observer moving at a constant velocity. Now let us appreciate its
different aspects.

It is a statement that certain relations—the laws of physics—have
an absolute significance, independent of the observer (provided he is
moving at constant velocity). One example of this general statement
is that the speed of light, which appears as a fundamental unit in
the basic laws of electromagnetism—Maxwell's equations—must be
the same for all such observers. The speed of light thus has an
absolute significance, independent of the observer.

At the same time, it is a statement that there are many equally
valid, but different, descriptions of the same physical reality. Observers moving at different constant velocities will see many things
differently. Although they will all agree on the speed of light, they
will in general see different colors in any particular light beam,
because of the Doppler shift. Also, of course, they will perceive differ-

ent velocities for things other than light beams—for instance, each will perceive himself to be at rest and his competitors in motion. So the color of a light beam, or the velocity of anything moving slower than light, is defined only <u>relative</u> to the observer. There are equally valid descriptions of reality, supplied by the other observers, in which the colors and velocities are different.

To capture the essence in an epigram: the relations expressed in the laws of physics are absolute, whereas the things they relate are relative.

Finally, the basic postulate is a postulate of <u>symmetry</u>. The essence of symmetry is just this: an object is said to be symmetric if it looks the same from several points of view. For example, a cube is symmetric because it looks the same to observers viewing it from six different positions; a sphere is still more symmetric because it looks the same to an infinite number of possible observers, viewing it from any angle. The basic postulate of relativity theory is that there is a very high degree of symmetry in physical laws—the laws must look the same from a very wide variety of "perspectives." Perhaps the best name of all for it would be neither of the two originally proposed—*relative* or *absolute*—but rather the postulate of *constant-velocity symmetry*.

Any statement of symmetry, in our very general sense, carries with it the dual aspects of relative and absolute. The requirement that an object look the same from different viewpoints implies <u>absolute</u> constraints on the structure of that object. At the same time, it means that several equally valid views of the object as a whole are available, according to which any particular aspect of it is viewed differently. The look of any particular aspect is therefore defined only <u>relative</u> to the observer.

After this discussion, we can illustrate in a graphic way the difference between a meaty statement of symmetry, such as the constant-velocity symmetry, and the trite statement "It's all relative to your point of view." Consider, as we have above, the case of shapes. To be sure, you can look at any shape from different viewpoints, and it will look like something or other, generally different for different viewpoints—this sort of relativity is hardly profound. The meat comes when the shape <u>appears the same</u> from <u>different</u> viewpoints. Then we say the shape is symmetric. The requirement that a shape be symmetric constrains the shape and at the same time liberates our description of it. The limiting case is perfect symmetry, where the same apparent shape is observed from all viewing angles. Then we both determine the object to be a sphere and ensure that we can describe it equivalently from many viewpoints.

23

Interchangeable Worlds

and lovers meander in prose and rhyme
trying to say—
*   for the thousandth time—*
what's easier done than said
                        —P. Hein, "Lilac Time" (1966)

We say objects are *symmetric* when they can be manipulated without being changed. Consider, for example, a stash of featureless gold bricks locked up in a vault at Fort Knox. Suppose that one morning the security guard looks into the vault and finds a mysterious note with the following message: "During the night I broke into your vault and turned one of the bricks around." Now, did someone actually break in and do this deed or did some philosophical joker merely slip a note through the cracks? (Or perhaps the joker projected it in, using a duplicator?) There is no way to tell by looking at the bricks, because a turned-around brick looks the same as one that hasn't been manipulated. The bricks are symmetric: turned-around bricks and bricks that have not been turned around are completely interchangeable.

When we speak of the symmetry of physical laws, we imply that the world can be manipulated without being changed. Now, this statement might appear to be an empty bit of cosmic chutzpah—after all, who's going to manipulate the world? And chutzpah it would be, if it weren't for two other profound features of the natural world. We have in mind, first, that the world is built from indistinguishable parts and, second, that physical laws are local. We have already discussed the radical identity of the building blocks of the world. That physical laws are local means simply that the influence of bodies far enough away seems to be very small.

It should be emphasized that this *locality* of physical laws is an empirical finding rather than a logical necessity. For instance, it is logically possible to imagine that the rate of chemical reactions might depend strongly on the phase of the moon—it just happens to be false. That it is false, is most fortunate for science. It would be extremely difficult to interpret the results of experiments if we really had to worry about the phase of the moon, not to speak of what the various planets or distant galaxies are doing. This is one reason astrology makes most scientists terribly uncomfortable. Similarly, it would be extremely difficult to do controlled experiments if telekinesis were for real, because then we'd have to take into account what everybody was thinking. Fortunately, however, it has proved possible to interpret the results of even the most delicate and sensitive experiments—for instance, experiments that measure the wavelength of spectral lines to parts in a billion or detect the flipping of one in a trillion proton spins in response to tiny magnetic fields—without taking these potential complications into account. In other words, such experiments are found to give the same results whatever the position of the planets and whatever your state of mind. You can isolate a laboratory from the outside world, and that's the essence of the proposition that physical laws are local.

The feasibility of creating, and isolating, identical systems gives a concrete operational meaning to the symmetry of physical laws. It isn't necessary to manipulate the whole world; it is enough to compare identical isolated systems.

For example, imagine comparing notes with a physicist doing experiments in an isolated rocket ship. She reports to us the results of every experiment she performs there: the spectra of atoms and molecules, the lifetimes of radioactive nuclei, the contents of her version of the Rosenfeld Tables—whatever. The essence of the symmetry of physical laws is that no matter how much she tells us about her world, there are important facts about it we will not be able to deduce.

One thing we will not be able to tell from her description of conditions inside her ship is where she is. That we can't is one way of stating the symmetry of physical laws under translation in space. Sealed laboratories that differ merely in position are interchangeable. You cannot, even from the most complete possible internal description, tell them apart. Similarly, we will not be able to tell, from her description, how she is oriented in space. She might be upside down or pointing any which way. Saying that you cannot tell is one way of saying that physical laws are symmetric under rotation of space. And finally in this vein, you will not be able to tell, from her complete description of her world, <u>how fast she is moving</u>. This is the statement of velocity symmetry, the essence of the special theory of relativity.

Of course, our rocket physicist could inform us of her position by reference to nearby stars, or of her orientation by sending us a gyroscope, or by many other possible methods. If there were no way at all for you to learn of these things, they would not be physically meaningful. The meaning of symmetry is that you cannot glean these physically meaningful facts purely from information, however extensive, she gives us about her <u>own isolated world</u>. She must refer us to something outside or send us a physical artifact. It reminds us of Piet Hein's grook, which we put at the top of the chapter.

Modern physics has uncovered many new symmetries and approximate symmetries of physical laws. A good example is colour symmetry, which is the statement that the different colour charges in the strong interaction are interchangeable.

To appreciate what this means, imagine that we asked our rocket friend about her theory of the strong interactions. "Oh, yes," she says, "I call it QTT, for quantum tint theory. According to my theory, protons are made of three little particles called aces. The aces come in three tints—beige, lavender, and chartreuse. The names of these tints are just a whimsy of mine; the different tints have nothing to do with light and are really more like three kinds of charge. . . ."

You will recognize, of course, that our friend is describing what we call QCD. Her aces are our quarks; her tints are our colours. But when we try to complete the dictionary, there's a hitch. How can we tell which tint corresponds to which colour? Is her beige our white, or our red or blue? It is not a meaningless question, because we could settle it by having her send us a beige quark. (Actually, the situation is a bit tricky, because one can't produce isolated quarks. But this complication is beside the point as far as our discussion of the idea of symmetry is concerned.) The question is whether we can tell in advance, before receiving it, what colour this beige quark will seem to us. And the principle of colour symmetry is simply that there is no way to anticipate.

We have stated the colour symmetry principle negatively, in the form of a principle of frustration. Its positive power may not be immediately obvious from this, but a little thought will make it evident.

For example, there are six different ways in which a quark of one tint (or colour) can emit a gluon and change into a quark of another tint (or colour). In the absence of a symmetry principle, the strength of each kind of coupling—or, in physical terms, the probability that such an emission will occur if the quark is accelerated—has no relation to the strength of the others. So there are six independent numbers specifying the couplings.

Now, we will argue that the principle of colour symmetry requires that all the couplings are equal. Suppose for a moment that, on the contrary, one transformation is more likely than the others. Let's say, to be definite, that our rocket physicist informs us that beige → lavender is the most likely. Then in our description of the strong interaction—since, after all, we are just describing the same facts in different words—there would also have to be a most likely transition. Let's say it was red → blue, to be definite. Clearly, then we could peg our friend's beige as our red and her lavender as our blue, just by comparing her description of her world with our observations of our own. In other words, the principle of colour symmetry would be violated. Turning it around, we see that colour symmetry requires that all the couplings are equal—a powerful simplification.

(It is possible to take the idea of colour symmetry an important step further. So far, we have assumed that the same laws of physics will still hold if one colour is interchanged with another. This is like saying that three particular directions in colour-space are equivalent. A more far-reaching assumption of symmetry is that the same laws of physics still hold if colour charges are continuously rotated into one another—in other words, that all directions in colour space are equivalent. This more complete symmetry brings with it the ultimate in simplifying power: when we demand full continuous colour symmetry, we find that the relative strengths of all the many possible couplings among the three colours of quarks and the gluon octet are completely fixed.)

The example of colour symmetry has shown how symmetries make it possible to simplify our description of the world. Symmetry, by allowing us imaginatively to manipulate the world, relates processes that otherwise seem utterly distinct. The remarkable thing is that nature seems to cooperate. The colour couplings <u>are</u> symmetric, as far as we can tell. As physicists have explored the microworld, symmetry has time and again proved a fruitful guide to guessing the form of physical laws.

24

More Perfect Worlds

Ah, Love! could Thou and I with Fate conspire
To grasp this sorry Scheme of Things entire,
Would not we shatter it to bits—and then
Re-mould it nearer to the Heart's Desire!
—Omar Khayyám, *Rubáiyát* (ca. 1100)

More and more, as our understanding of fundamental physics has progressed, we have come to ascribe the rich texture of reality to ideally simple laws acting in a complex environment.

Consider, for example, ideas about space and motion. Even sophisticated Greek and medieval philosophers did not recognize the symmetry among different directions in space. After all, things do fall down rather than sideways, and it is much easier to walk along the surface of the earth than to bore into it. It requires considerable abstraction and imagination to realize that the three dimensions of space are equivalent. Nor, for similar reasons, did our ancestors recognize the principle of inertia, that in the absence of external forces bodies travel with constant velocity in a straight line. We learn in childhood, and experience every day, that it takes continuous effort to run or walk at a constant velocity, despite the principle of inertia. It took hundreds of years and the daring genius of men like Descartes, Galileo, and Newton before people learned to blame the earth, its gravitational field, and the friction of its atmosphere for hiding a deeper symmetry and simplicity.

Descartes, Galileo, and Newton realized that it was a good idea to formulate the basic laws of physics as they hold in empty space. The local environment could then be added in as a complication and blamed for hiding the underlying symmetry and simplicity of physical law. In the recent quest for more and more symmetry, physicists have gone a step further. Nowadays they are no longer content with real empty space, such as could be found far from the earth, and clear of other "matter." They have become convinced that so-called empty space is a complicated environment. They have, in other words, learned to "blame the vacuum" for many of the complications in the observed laws of nature. It seems that unavoidable background fields permeate even what we normally call perfect vacuum—space as empty as we can ever find it. These fields hide the full simplicity and symmetry of physical laws from us. We must learn to imagine more perfect worlds, emptier vacuums.

To make an analogy, imagine for a moment a magnetic universe —a universe everywhere pervaded by a constant magnetic field. In this universe, the perfect symmetry between different directions in space would be spoiled—indeed, we could orient ourselves in empty space by using a compass. And the laws of motion for charged particles would be complicated, because the magnetic field would cause them to spiral around. Now, the inhabitants of this magnetic universe might at first describe its laws just as they found them. Then they would have an asymmetric, complicated formulation of physical laws. Eventually, though, they would surely find it pleasant to imagine a world free of the pestiferous field, a world with beautifully symmetric and simple laws. They would learn to say that the "real" laws of physics are the symmetric and beautiful ones and that these laws are hidden from view by a vacuum that is not empty enough.

Much like the inhabitants of this imaginary magnetic world, physicists have come to feel that ordinary empty space is just not empty enough.

Let us trace the path that leads to this extraordinary conclusion. We recall that electrons have *spin,* which for present purposes we can think of as simply meaning that they are little gyros, always rotating at the same speed around an internal axis. We classify moving electrons as left- or right-handed, depending on how they're rotating. Imagine you are riding an electron, sitting at the front and looking in the direction of motion (as if you were driving a truck). Then we will say the electron is *left-handed* if its rotation carries your head toward your left hand and *right-handed* in the opposite case. To illustrate this definition: if the earth as a whole were moving north, it would be right-handed. Now, it was a remarkable discovery, made in the 1950s, that electrons emitted in weak interactions (such as neutron decay) are almost always left-handed. Almost always—but not quite always. It seems that nature is trying to tell us something, but the message is coming through slightly garbled.

Actually, some garbling of this particular message is inevitable, because the distinction between left-handed and right-handed electrons is not completely sharp. You see, to define the handedness of the electron, we have to decide in which direction it is moving. Suppose, for instance, that the electron appears to us to be moving north. To an observer moving north even faster than the electron, the same electron will appear to be moving south. So an electron that appears left-handed to us will appear right-handed to the rapidly moving observer. The principle of relativity tells us that observers moving with any constant velocity must see the same laws of physics, but "left-handed electron" is not a concept all such observers agree on. So, if the principle of relativity is correct, it cannot be strictly true that only left-handed electrons emerge from weak decays.

On the other hand, in a world where electrons had zero mass this problem in identifying left-handed electrons would not arise. In such a world, electrons would always travel at the speed of light, and (since this is the limiting velocity) we would not have to worry about observers possibly moving more rapidly. Even observers moving with different velocities would be able to agree on the handedness of electrons. In this case, it would be a possible law of physics—consistent with the principle of relativity—that only left-handed electrons are emitted in weak decays.

Nature has apparently given us a hint, in the weak interaction, that our world is in some sense close to this simpler one. It is an elegant idea, and proves fruitful, to try to blame the vacuum for the difference. So let's consider the possibility that there is a background field permeating all space, which is responsible for giving the electron mass. In other words, we are proposing that an unavoidable interaction of electrons with the background field slows them down.

(Wait a minute—if electrons are slowed down by their interaction with this pervasive field, aren't we in danger of repealing the principle of inertia? No. The background field slows electrons down by comparison with the speed they would have had if the field weren't present, but they don't slow further as time progresses.)

The mass-generating field is known as the Higgs field, after Peter Higgs, a Scottish physicist who discovered some of the advantages of blaming the vacuum for the masses of particles. An intriguing idea—but is it true? The most direct tests of the idea would be to compare physics with and without the background Higgs field. Unfortunately, as we will discuss below, it is not a practical proposition to turn the background Higgs field off over large regions of space in our laboratories. But some spectacular indirect tests may soon be possible. We may soon be able to break off and observe tiny chips of the Higgs field—an exciting prospect.

This hasn't been done, but even without direct experimental evidence most physicists are convinced that the concept of the Higgs field is here to stay, for it allows us to imagine more perfect worlds and relate them systematically to the world we live in. Let's see how.

The full power of the Higgs field idea becomes apparent when we widen our horizons to consider other particles besides electrons. The weak interaction tends to favor the left-handed form not only of electrons but of all fundamental fermions as well: the various kinds of quarks, electrons, muons, tau leptons, and neutrinos. In other words, it is more likely that these particles are emitted left-handed than right-handed. (On the other hand—ha, ha—the weak interaction favors the right-handed form of antiparticles.) As we discussed in regard to electrons, this rule is not exact. It can't be, at least not if the theory of relativity is correct. However, it is found that the rule is broken in a very particular and structured way.

The way the rule is broken is most easily described by first imagining a more perfect world, in which the background Higgs field vanishes. In this imaginary world, all the fundamental particles we just mentioned will be massless. So in this world, it <u>can</u> be an exact law that only the left-handed particles participate in the weak interaction. Now, suppose that our world differs from this imaginary one only in the presence of an all-pervasive background Higgs field. While the interaction of particles with this background field slows them down (gives them mass), it does not alter their weak interactions. The weak interactions still favor the left-handed form. The background Higgs field can occasionally cause the spin to flip, thus changing the left-handed particle into a right-handed one. But it can be calculated that for a rapidly moving particle this happens only rarely. From this perspective, the approximate rule that the weak interactions tend to favor left-handed particles is seen to encode an <u>exact</u> law whose operation is partially obscured by the background Higgs field.

The great advantage of introducing the Higgs field is that it allows us to imagine how much more beautiful physical laws would be in its absence. It is a most convenient scapegoat.

The background Higgs field must have very accurately the same value throughout the universe. After all, we know—from the fact that the light from distant galaxies contains the same spectral lines we find on Earth—that electrons have the same mass throughout the universe. So if electrons are getting their mass from the Higgs field, this field had better have the same strength everywhere. What is the meaning of this all-pervasive field, which exists with no apparent source? Why is it there?

Physicists believe it arises as the result of an instability, similar to instabilities encountered in less exotic situations. Consider, for example, a lump of iron oxide. The fundamental interactions of its electrons and nuclei do not recognize any special direction in space; they are perfectly symmetric. And yet, our lump of iron oxide will organize itself into a magnet. Electrons throughout the lump will all align their spin axes in some particular direction, pointing toward the poles of the magnet. Once this occurs, of course, the perfect symmetry between the directions of space is destroyed. If there were intelligent beings embedded in iron oxide, they would find themselves occupying something resembling the imaginary magnetic universe we discussed before.

How can symmetric laws give rise to an asymmetric result? It is a little subtle, but not terribly difficult to understand. There are forces between the spins, which tend to align them with <u>each other</u>. Now, if it is energetically favorable for all the electrons to align their spins, they must make a collective choice, and all come to agree on just one direction. And so the forces that line spins up, although they do not themselves distinguish any particular direction, compel the spins to pick one out. Symmetric forces enforce an asymmetric solution. <u>The laws of the world are more symmetric than any stable realization of them</u>. This is the great lesson of the magnet.

Although the relevant forces and interactions are not yet known in detail, physicists strongly suspect that a similar effect is responsible for the background Higgs field which permeates our universe. In other words, the answer to the question "Why isn't our vacuum more empty?" is that emptiness is unstable.

The existence of the background Higgs field permeating all space, which we blame for spoiling the beauty of our equations, is a standing challenge. Can we get rid of it? Or, to be more modest (and prudent), could we at least significantly modify it? Unfortunately, the energies required to do this over any reasonably large volume are calculated to be far beyond what we can muster. A modest step in this direction may, however, be possible. Since the Higgs field is supposed to be responsible for particle masses, heavy particles clearly couple to it most efficiently. And so we expect that, in the immediate neighborhood of heavy particles, the Higgs field will be significantly changed from its value in the vacuum. (Just as the electric field is significantly changed from its value in vacuum—namely, zero—in the immediate neighborhood of highly charged nuclei.) For instance, in the neighborhood of toponium, a bound state of t quarks and t̄ antiquarks, or in the neighborhood of a Z boson—to mention two heavy objects that accelerators should be able to produce in great abundance in the near future—the Higgs field is changed. These objects make a little dimple, if you like, in this universal sounding board. Now, both toponium and Z bosons are unstable. When they decay into lighter particles, the disturbance in the Higgs field (the dimple in the sounding board) is left behind with nothing to maintain it. The deformation will disappear, and the energy associated with it must take on some other form. One possibility is that it will become a traveling disturbance in the Higgs field itself—a packet of energy propagating outward, a wave in the sounding board, or, in other words, what we call a *Higgs particle*.

The Higgs particle is to the pervasive mass-generating Higgs field what the photon is to electromagnetic fields. In view of this relationship, we are tempted to call it a *chip off the old vacuum*. Unfortunately, no one knows how to predict the mass of the Higgs particles precisely, so we cannot be certain that they will observed in the near future. However, it is quite possible—even likely—that they will be observed in the next five years or so. That is when accelerators capable of making lots of toponium or Z bosons are due to come on-line. If Higgs particles are found, our faith in a more perfect, symmetric, and beautiful world will have been rewarded. If all goes well, we will be able to check in detail whether the Higgs field really couples as advertised, by examining how little chips of it behave. For one thing, we can see if it really couples most strongly to heavy particles, as predicted. The Higgs particle should decay much more

often into a bottom quark–bottom antiquark pair than into an electron–positron pair, for example, because bottom quarks are much
heavier than electrons.

Our present ability to produce conditions extreme enough to alter
the background Higgs field is limited, but it appears quite likely that
in its early history the universe was hot enough to <u>melt</u> it away.

Think again of our lump of iron oxide, which makes a magnet. If
you heat the lump up past 893 degrees Kelvin, it will cease to be a
magnet. This behavior of our lump is the result of another battle
between energy and entropy. The lowest energy is attained when all
spins are aligned—and the lump is magnetic. By contrast, the highest entropy is attained when the spins are oriented independently,
since then there are many more possibilities. At low temperatures,
energy is the dominant consideration, but as the temperature rises,
the randomizing influence of entropy wins out. To put it another
way, forces that can hold the spins in line when the spins are calm
become incapable of restraining them when they get highly agitated.

Similarly, the forces that currently make it favorable for the
Higgs field to take the same value everywhere in space will not be
sufficient to do so at arbitrarily high temperatures. If this is right,
then early in the history of the universe the fundamental particles
we are familiar with were all massless, and the equations describing
the world were simpler and more symmetric.

Could any relic from the brief symmetric phase in the early history of the universe, when it was free of the Higgs background, or
from the possibly violent transition from the symmetric Higgs-free
phase in the early history of the universe to the present phase, have
survived? This fascinating question is a topic of current research.
Physicists will be in a much better position to answer it after they
know the mass and properties of Higgs particles.

Once we learn to imagine more perfect worlds, and come to realize
how easy it is to blame the vacuum for the lack of perfect symmetry
in physical laws, it seems a pity to stop. Might there be other, even
better-hidden symmetries, broken by still more obscure Higgs-like
fields? There are powerful hints that this is true—read on.

25

A Suggested Unity

*An insanity as enormous, as complex, as the one
around me had to be planned. I've found the plan!*
—R. Heinlein, "They" (1948)

One fascinating problem under study today is whether the apparent diversity of the three interactions—strong, weak, electromagnetic—that dominate the microworld is real or illusory. Can the differences be explained and our description of nature unified? Guided by intuitions of symmetry, a few facts, and analogies with known interactions, physicists try to create models describing the real world. Then these models get judged, according to how well they match that real world—and by aesthetic standards as well.

There now exist very realistic models of the strong, electromagnetic, and weak interactions. QCD and QED, we think, also get high marks for internal beauty. The standard model of the weak interaction is less aesthetically pleasing—a bit more lopsided and clearly unfinished.

In this chapter, we shall describe a possible unity among all three of these interactions. *Grand unified theories* unite all three interactions in an extraordinarily ingenious and beautiful way. The ideas are bold, and they have been tested against reality at only a few points. Their realism remains in doubt. Yet they are so attractive, and follow so logically from well-tested theories, that they seem likely to contain important elements of truth and certain to inspire us for a long time to come.

To understand the nature, and indeed the possibility, of the suggested unity, it is first necessary to explore the modern theory of the weak interaction more deeply.

The weak interactions are marked by their power to transform the particles they act on. We have met a fundamental theory of transformations before: QCD, the colour theory of the strong interaction. Can the weak interaction, whose transforming power is much more obvious, be described by a similar, colour-full theory—a theory of transformation among colour charges?

We are trying to identify transformations of some new kinds of colour with the transformations observed to occur in the weak interactions. Remember, the rules of quantum mechanics and relativity are so powerful, so restrictive, that to give a recipe for reality it is necessary only to add to these rules a list of the ingredients in the universal ocean and of the reactions among these ingredients.

We begin by neutralizing—imaginatively, at least—one especially troublesome ingredient. The Higgs background complicates things by blurring the distinction between left- and right-handed particles, a dichotomy absolutely fundamental in the weak interactions. So let us begin by visiting a more perfect imaginary world, where the Higgs background has been turned off. Once we understand the simpler world, we will find it relatively easy to understand how the presence of the Higgs background changes things.

In the Higgs-free world, we expect to find that the rule that only left-handed particles appear in weak decays will be not just approximate but absolute. The weak colours, therefore, will be assigned to left-handed particles only; their right-handed antiparticles get the corresponding anticolours.

The next step is to decide how many weak colours to start with and how to assign them to the known particles.

To begin, let's look once more at our familiar paradigm of the weak interaction, neutron decay:

$$n \rightarrow p + e + \text{\reflectbox{$\mathbb{V}$}}$$

"Behind the scenes" of this weak decay, a d quark is being changed into a u quark. We are going to call this a weak colour transformation—from (why not?) yellow to green. That means assigning a unit of yellow colour charge to each left-handed d quark and a unit of green colour charge to each left-handed u quark.

To mediate such a transformation, we need a yellow → green gluon (called, for obscure reasons, a W boson). Finally, to complete the decay, the virtual gluon must materialize as an electron and an antineutrino. These ideas are summed up in figure 25.1.

1.		**d**	**→**	**u**	**+**	**W**
		(down quark)	→	(up quark)	+	W boson
Electric charge:		$(-\frac{1}{3})$	=	$(+\frac{2}{3})$	+	(-1)
Green colour charge:		0	=	$(+1)$	+	(-1)
Yellow colour charge:		$(+1)$	=	0	+	$(+1)$

2.	**W**	**→**	**ν**	**+**	**e**
	W boson	→	antineutrino	+	electron
Electric charge:	(-1)	=	0	+	(-1)
Green colour charge:	(-1)	=	(-1)	+	0
Yellow colour charge:	$(+1)$	=	0	+	$(+1)$

Figure 25.1.
Neutron decay in two steps.

As you can see, the colours flow smoothly if the left-handed electron "takes" the gluon's unit of yellow charge, and the right-handed antineutrino carries off a negative unit of green charge. (Its opposite number, the left-handed neutrino, therefore will have a positive unit of green charge.)

So, with just two new colours and the associated transforming gluon, we can build a lifelike model of neutron decay. It turns out that two colours suffice to describe all the other weak transformations as well—muon decay, K-meson decay, charmed-quark decay, and so on.

The colour language is a convenient way of accounting for weak transformations we already know about—and *accounting* is a good word for this bookkeeping style of explanation. But the symmetry of weak colour is more than bookkeeping. It carries within it a startling new phenomenon.

As in QCD, so in the colour theory of the weak interaction there is an additional type of "gluon" that couples to, but does not change, the new colours. This weak interaction gluon, called the Z boson, plays an important role in the description of neutrino behavior. Neutrino interactions mediated by W bosons are relatively easy to study, because the W boson changes a neutrino into an electron, whose charge and mass make it easy to detect. In the so-called *neutral current* interactions mediated by Z bosons, however, the neutrino remains a neutrino and merely deposits some of its energy and momentum. Since neutrinos leave no tracks—being neutral and in general interacting very feebly—what's actually observed is "nothing in–nothing out," merely an occasional deposit of energy with no visible cause. Before saying that such an event is due to an interesting neutrino interaction, experimenters must carefully eliminate all other possible explanations—for instance, stray neutrons, photons, or cosmic rays. But by taking on the challenge of detecting neutral currents, experimenters performed a great service; they demonstrated that the Z boson, a particle predicted only by the colour theory of the weak interaction, did in fact exist.

And so, in our imaginary Higgs-free world, the true nature of the weak interaction has been revealed. It is a theory of transformations among symmetric colour charges, much like QCD. Two new colours are involved; distinct from the three colours that appeared in the strong interactions.

Before taking leave of this imaginary world, let us pay tribute to its symmetry. When we first introduced the idea of switching off the Higgs background, we did so to make the laws of physics look simpler. Now you can see how removing the Higgs background also makes the laws look more symmetric, by allowing us to interchange the two weak colours. Just as the red, white, and blue versions of any quark—for example, red, white, and blue u quarks—came to look like inseparable aspects of a single reality, so do the pairs of particles, yellow and green, made perfectly equivalent by the interchange of the weak colours. For instance, in our imaginary Higgs-off world, where W bosons are massless and omnipresent, the left-handed electron and its neutrino exchange their identities in a constant flip-flop of weak colours. They are interchangeable. And so the existence of these two particles, which seem very different from one another, completes a hidden harmony.

This imaginary world is too perfect, too symmetric, to match reality. To have a model we can try to identify with reality, we must imaginatively switch the Higgs field on again. How does this change our imaginary world of perfect symmetry? One change, of course, is that the familiar particles acquire mass, spoiling the symmetry between weak colours. Left-handed electrons are no longer interchangeable with left-handed neutrinos. Not only do they have different masses, but also the very identification of "left-handed" electrons becomes ambiguous.

From a practical point of view, an even more fundamental change is that the W and Z bosons acquire large masses. This makes it harder for quantal fluctuations to create significant changes in the W and Z fields; in other words, it makes exchange of virtual W and Z particles rare. This enfeebling effect of the large mass of the W and Z bosons is mainly responsible for making the weak interactions weak. The large mass of W and Z bosons makes it hard not only for quantal fluctuations to create the virtual version of these particles but also for experimenters to produce the real version. Powerful accelerators are needed to concentrate sufficient energy. Even though the existence and properties of these particles had been inferred from their role in weak interactions a decade earlier, it was not until 1983 that they were actually produced.

We have identified the observed strong interactions with transformations of red, white, and blue colour charges among themselves, and the observed weak interactions with transformations of green and yellow colour charges into one another. They appear analogous but still occupy two separate worlds. The bold idea of *grand unified theories* is that both are aspects of a larger whole. According to this idea, the basic rule, from which strong, weak, and electromagnetic interactions all flow, is that "gluons" exist to make <u>all possible</u> transformations among the five colours.

Another statement of the grand unification idea is the doctrine that <u>all colours are created equal</u>. Not only are the three strong colours strictly interchangeable with one another, and the two weak ones imaginatively so (when the Higgs background is removed); the new idea is that even <u>strong and weak colours can be interchanged with one another</u>, once an *ultra-Higgs* background is removed. The promise of grand unified theories is that our world would be revealed as perfectly symmetric among all five colours, if only pestiferous Higgs and ultra-Higgs backgrounds could be turned off.

According to this theory, just as the Higgs background operates to spoil the symmetry between green and yellow colour charges, the ultra-Higgs background field hides from us the symmetry between strong and weak colour charges. This ultra-Higgs field operates to give enormous masses to the particles, called X bosons, that mediate strong-weak colour transformations, just as the ordinary Higgs field gives mass to the W and Z bosons.

Grand unified theories promise a breathtaking simplification and symmetry among the interactions that govern the microworld—the strong, weak, and electromagnetic interactions. But before we can take this attractive idea quite seriously, a fundamental question must be addressed: If these interactions are "really" all alike, why do they appear so different?

You probably can anticipate that we'll blame the vacuum for this deceptive appearance. To a certain extent we've done that already, by holding the Higgs background responsible for the mass of the W and Z bosons. However, the heaviness of the W and Z bosons does not fully explain the difference between their interactions and those of their putative equals, the gluon octet and the photon. Nor is it obvious how the gluon octet and the photon themselves can in any sense be equals. In fact, it seems absurd on the face of it—even after correcting for the enfeebling effect of mass, we find a wild mismatch in coupling strengths for the three types of interactions. The coupling of the gluon octet to the strong colour charges remains far stronger than that of the photon to electric charge, or of the W and Z to weak colour charges.

Is grand unification therefore exposed as a seductive illusion? Not at all; in fact, it is precisely at this most vulnerable point, in comparing the power of the different interactions, that the idea of grand unification scores its profoundest triumph.

The key to this triumph is asymptotic freedom, the realization that the strong interaction gets weaker at high energies or short distances. We should expect that the relative power of the different interactions will become equal only at energies large enough to "melt" the ultra-Higgs field that hides from us the ultimate in symmetry. Only at energies high enough to permit the interchange of strong and weak colours must the couplings necessarily be equal. But at such energies the strong interaction will have become much feebler, and so it <u>will</u> have a chance to match the others.

Turning the logic around, we should be able to start with equal couplings for all the interactions at ultrashort distances and calculate the changes in the effective couplings at larger distances due to screening and antiscreening by virtual particles. If grand unified theories are correct, we ought to be able to derive the relative power of the strong, weak, and electromagnetic interactions at accessible energies from their presumed equality at much higher energies.

When this is attempted, a wonderful result emerges. It is shown, in the form first calculated by Howard Georgi, Helen Quinn, and Steven Weinberg, in figure 25.2. The couplings of strong-interaction gluons decrease, those of the W bosons stay roughly constant, and those of the photon increase at short distances—so they all tend to converge, as desired.

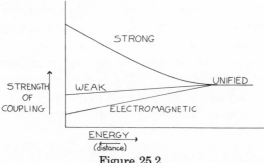

Figure 25.2

We're talking not just about an approximate convergence but about a real numerical fit between experiment and theory. There are three couplings to compare. By comparing any two of them, we can determine the scale of unification. In other words, we can find out to what extremes of tiny distance or enormous energy we must proceed before these two couplings become precisely equal. Having done this, we have nothing left to juggle. The power of the third interaction is then fixed; its value becomes a critical numerical prediction of the grand unified theory.

Remarkably, the prediction fits reality to within a few percent. This is a most impressive indication that the idea of grand unification is on the right track and that the ideas of virtual particles, screening, and asymptotic freedom—in other words, applied relativity and quantum mechanics—are almost surely valid down to distances thousands of trillions times smaller than those at which they were originally discovered and tested.

The general idea of unifying the interactions is, we have seen, not only attractive but also successful. But a scientific idea must rest upon specific predictions; grand theories must depend on a basis of realistic detail. And so model-builders happily take up the challenge of painting candidate worlds in five colours, hoping to find one that resembles our own in realistic detail.

To put it more plainly, the task of model-builders is to find a way of assigning colour charge to the known particles that is symmetric and yet provides a correct description of their interactions. Since the requirement of symmetry among five colours is very restrictive, success in this endeavor is far from automatic. The results, however, are spectacular.

An astonishingly simple and symmetric pattern emerges, linking colours together for all the known particles—everything seems to fit together just right. The centerpiece of grand unification is encoded in the accompanying table, which we call the Unification Table. Like the rainbow or the spectrum of hydrogen, this table is a mighty symbol of beauty and order in the physical world. You should pause a moment simply to look at it. . . .

Red	White	Blue	Green	Yellow	Particle Name	Strong	Weak	Electric Charge
+	+	+	+	−	**v** neutrino	no	yes; with **e**	0
+	+	+	−	+	**e** electron	no	yes, with **v**	−1
+	+	−	+	+	**d** anti-down quark	B	no	+1/3
+	−	+	+	+		W		
−	+	+	+	+		R		
+	+	−	−	−	**u** anti-up quark	B	no	−2/3
+	−	+	−	−		W		
−	+	+	−	−		R		
+	−	−	+	−	**u** up quark	R	yes; with **d**	+2/3
−	+	−	+	−		W		
−	−	+	+	−		B		
+	−	−	−	+	**d** down quark	R	yes; with **u**	−1/3
−	+	−	−	+		W		
−	−	+	−	+		B		
−	−	−	+	+	**e** antielectron (positron)	no	no	+1
−	−	−	−	−	**N** left-handed antineutrino	no	no	0

And now let us briefly explore its meaning. The first five columns represent the strong and weak colours. They contain all possible combinations of five + and − signs, with only one restriction: that there must be an even number of + signs or, equivalently, an odd number of − signs. Associated with each of these combinations is the name of a left-handed particle (**boldface letters**) or antiparticle (outline letters). We can generate the right-handed antiparticle for each of these southpaws by switching all five colour signs. For example, the left-handed electron's colours are given as $+ + + - +$; the corresponding colours for its antiparticle (a right-handed positron) would be $- - - + -$.

Each particle's strong and weak and electromagnetic interactions are determined by its colour pattern. (To learn how to figure out electric charge from colour charge, and other arcana, see the coda at the end of the chapter.) So the first five columns, with their codelike pattern of $+$ signs and $-$ signs, embody in shorthand all the information in the chart—in other words, they tell us everything there is to know about the strong, electromagnetic, and weak properties of the fundamental particles. (The N particle has no trace of colour charge or of electric charge. In fact, the N is so devoid of any possibilities for affecting the real world that we shall mostly ignore it.)

A simple rule prescribes the colour-transforming behavior of both strong and weak gluons. It is this: transforming gluons act by switching signs $+$ to $-$ in one column of the table, and $-$ to $+$ in another. Some of these gluons are members of the octet of strong colour gluons; other are the W bosons; still others are the very heavy X bosons mentioned earlier, whose stunning potential you will see in a moment. Consider a concrete example: the gluon that flips $+$ signs in the first column and $-$ signs in the second is capable of transforming $+ - - + -$ into $- + - + -$. That is, it turns a red u quark into a white one. It can also change $+ - + + +$ (an antiwhite anti-d quark) into $- + + + +$ (an antired anti-d quark). On the other hand, this gluon could not transform any particle that already had a $-$ sign in the first column, or a $+$ sign in the second. You can convince yourself that this gluon transforms everything in the table just as does the familiar, bona fide red $\rightarrow$ white gluon of the strong interaction. Similarly, the transformations that flip signs in the green and yellow columns match those induced by the W bosons. In short, the picture of the world summarized in the Unification Table represents a remarkable triumph of model-building. Diverse interactions of disparate particles flow from a single simple, systematic rule.

In the table—and in nature—we find (leaving aside N) fifteen fundamental fermions, with diverse strong, weak, and electromagnetic charges. Some of these are not really independent. They are so closely related by symmetry transformations that they are, so to speak, no more than different faces of the same cube. For instance, the red u quark is inseparable from the white u quark, because there is a symmetry relating these two colours. If we group together the quarks that differ only by a transformation of strong colour, the number of distinct fundamental entities is reduced to seven. If we take the next step and group together the pairs that differ only by a transformation of weak colour, this number is reduced to five.

At this point, we have used up all the unifying power of the strong and weak transforming principles. But if we admit the additional transformations of grand unification, then all particles with four + signs become interchangeable, as do all the particles with two + signs. The number of fundamental, noninterchangeable entities is only two. So the multiplicity of diversely interacting fundamental particles is cut back to mere duality by the more inclusive symmetry.

Can we take the final step, and reduce two to one? It is mathematically possible, and therefore almost irresistible, to take the symmetry one step further and unify the fermions completely. The wider symmetry requires a whole new class of transforming gluons, able to flip signs + to − in both of two columns. In this scheme, the fifteen fundamental fermions, and also **N**, are united to make one inseparable whole.

Why are the laws of physics so symmetrical? It is not clear whether this is a proper scientific question or whether a proper scientific answer to it can ever be given. If you tell me that the laws of physics are so symmetric because of Grand Principle Omega, I can always ask you why Grand Principle Omega is true. Nevertheless, the form of the Unification Table provokes a wild thought—call it a dream, a nightmare, or a speculation as you please—regarding the symmetry of the laws.

When we view nature stripped to essentials, so to speak, in this Unification Table, what we see is . . . a five-bit register. Each particle can be specified by using just five + or − signs—or equivalently by any set of five binary choices, like "0-or-1" or "on-or-off." It is in just this form that data is stored and manipulated within a digital computer. In computer jargon, one binary choice conveys one "bit" of information. Each particle, then, can be specified by a five-bit word and stored in a five-bit register.

It's eerie that even the odd restriction on how many minus signs are allowed is reminiscent of a trick used in computers, to detect errors in transmission. You see, if all allowed words must have an odd number of minus signs, any single error in transmission of a word can be detected. If a single + is changed to a −, or vice versa, the botched result will not be an allowed word.

So the different particles are like different possible states of a five-bit register. The state of the world at every point of space-time is given by the values of the fields associated with these particles—that is, by a huge net of interconnected registers. The uniformity of physical laws or, in other words, their symmetry in time and space means that the rules for computing are fixed once and for all and are the same throughout the net. The colour symmetry means that the rules are particularly simple and do not distinguish between different locations in the registers. And the locality of physical laws means that the net is connected in an especially simple way: each register communicates with only a few others—its near neighbors.

These features of uniformity and locality, in a computer, would help make for flexibility and ease of programming. They imply that one can construct modules "occupying different regions of space" (that is, sharing no connections), which all speak in a common language. Yet, because of locality, such modules are quasi-independent—communicating only at common boundaries—and interference among messages is avoided.

We therefore come to suspect, from its design, that our world just might be an intricate program working itself out on a gigantic computing machine. This form of paranoia may seem extravagant and, of course, doesn't get to the bottom of explaining the world. We would still need to understand the principles on which the computer was built. For instance, if it is made of silicon, why do the electrons in that silicon obey the laws of physics—are they perhaps fantasies in yet another computer??

Nevertheless, it may not be completely useless to follow up on this suspicion. First of all, it does begin to address the great "why" questions in a rational, if possibly mistaken, way. Second, it suggests a fascinating new sort of question: How would errors in the workings of the computer show up? Could we look for them systematically, by experiments, and thus put the idea of an underlying computing machine to a scientific test?

Finally and perhaps most important, thinking along these lines will help prepare us for the day when we—or, more likely, our distant descendants—will develop the machinery and cleverness to begin to program worlds ourselves (and watch—with what feelings?—as the inhabitants of those worlds come to start suspecting . . .).

Coda: About the Table

Here we discuss the connection between the Unification Table and the properties of particles in more detail. Nothing in the rest of the book depends on this, rather technical, coda. In order to make the discussion self-contained, we begin by recalling a paragraph from the text.

The first five columns represent the strong and weak colours. They contain all possible combinations of five $+$ and $-$ signs, with only one restriction: that there must be an even number of $+$ signs or, equivalently, an odd number of $-$ signs. Associated with each of these combinations is the name of a left-handed particle (**boldface letters**) or antiparticle (outline letters). We can generate the right-handed antiparticle for each of these southpaws by switching all five colour signs. For example, the left-handed electron's colours are given as $+++-+$; the corresponding colours for its antiparticle (a right-handed positron) would be $---+-$.

The link between the particle and all the signs is simply this: the particle carries half a unit of the colour charge if a $+$ sign appears in the column for that charge, and minus half a unit of the colour charge if a $-$ sign appears. For instance, the left-handed electron has half a unit each of red, white, blue, and yellow charge, and minus half a unit of green charge.

The strong, weak, and electromagnetic interactions of the particles are governed by these colour charges. In the next column of the table, labeled "strong," the strong colour charges are displayed in a cleaned-up form. You might remember there was a rule that played an important role in our understanding of how the strong colour charge of protons is neutralized, according to which the simultaneous presence of the three different strong colour charges adds up to no strong colour charge at all. As an example of this rule—let's call it the *bleaching* rule—the strong colour charge of the electron that follows from our table is zero. It had better be, since electrons don't participate in the strong interaction—otherwise, what we called an electron in our table just wouldn't be an electron.

A slightly more intricate use of the bleaching rule occurs for the u quark. The $+ - -$ in the first three columns means one-half unit of red charge and minus one-half unit of blue and white charge. But minus one-half unit of blue and white charge can be replaced, according to the bleaching rule, by plus one-half unit of red charge. Altogether, then, the $+ - -$ in the first three columns can be interpreted simply as one full unit of red charge. After these examples, it should now be easy for you to check all the entries in the "strong" column.

Next we move to the "weak" colours. There is a bleaching rule for weak colours also, so any particle with either $+ +$ or $- -$ in the green and yellow columns has no net weak colour. On the other hand, the $+ -$ indicates that the particle carries one unit of green charge and the $- +$ that it carries one unit of yellow charge. Particles that differ only in the weak colours can be transformed into one another by emission of a W boson, and you can check that in this way the table gives us precisely the basic weak interactions we want.

Finally, there is the "electric charge" column. The electric charge is given by a rather complicated rule. Skip over it if you like, but please don't let its ungainly appearance put you off—it really doesn't appear quite so ungainly in the full mathematical context of the theory. Without further apology: the rule is that the electric charge of a particle is obtained by taking $-1/3$ times the sum of the red, white, and blue charges and then adding the green charge. For example, the electron has electric charge $-1/3 \times (1/2 + 1/2 + 1/2) + (-1/2) = -1$.

The strong, weak, and electromagnetic charges of all of these fermions have therefore been derived on a uniform basis, from simple though abstract principles. In the symmetry and inevitably of the Unification Table, the grandeur of the grand unified theories is manifested.

26

Salt Mine

If you don't make mistakes, you're not working on
hard enough problems. And that's a big mistake.
—F. Wilczek, Unsolicited Advice to Graduate
Students (1974–?)

Inside the Morton salt mine near Cleveland, Ohio, a thousand feet beneath the surface of the earth, there lies a huge vat, ten meters on a side, of the clearest water in the world. Surrounding it, in the cold and dark, automated eyes maintain a tireless vigil, awaiting rare flashes that will signal the decay of individual protons.

Proton decay—if it occurs at all—must be an extremely rare process. But one of the most spectacular and most experimentally accessible results of grand unification is that protons <u>do</u> decay. The predicted process, shown here in figure 26.1, is mediated by gluons that change strong into weak colours.

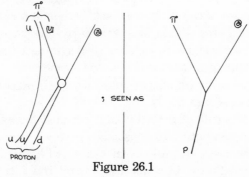

Figure 26.1

It's easy and fun to see how the possibility of proton decay is encoded in the Unification Table. The table shows that when a red d quark emits the virtual gluon carrying a unit of red charge and a negative unit of yellow charge—that is, the gluon that flips + to − in the first column and − to + in the fourth column—the d quark changes into a positron. This same virtual gluon, when absorbed by a blue u quark, changes it into an antiwhite anti-u quark. (Clearly, English is not well adapted to this sort of symbol manipulation—it's "easier done than said.") And so we have derived the process shown in the left half of the figure. If you take a look at the right half of the figure, you'll see how this process leads to proton decay.

No one has yet seen a real proton decay. Even crude calculations suggest that its lifetime is long. If, for example, protons decayed in human bone, the energy released would increase the incidence of cancer. From this consideration, it is possible to show that the (average) lifetime of protons must be greater than 10^{16} years. If the protons within Jupiter decayed, the liberated energy would contribute to the luminosity of the planet. The luminosity would be greater than is actually observed, if the lifetime were less than 10^{18} years. Remember, for comparison, that the lifetime of the universe is only 10^{10} years, so that at most only a very small proportion of the protons in the universe have decayed since the big bang.

The next question is, What rate of decay should we expect, if the new X gluons predicted by grand unified theories work as advertised? The answer rests on a chain of reasoning that goes as follows:

The rate of proton decay depends on the mass of the mediating gluons. The larger their mass, the rarer the quantum fluctuations that bring them to life, and hence the longer the lifetime of the proton. The mass of the X gluons is set by their interaction with the ultra-Higgs background. The size of the ultra-Higgs background is in turn related to the degree to which perfect symmetry between strong and weak colour interactions is violated—concretely, to the disparity between the strength of these interactions. And this, at last, can be measured experimentally.

To predict the rate of proton decay, we must follow this chain in reverse. The computations basically involve turning the idea sketched as figure 25.2 into a real graph, with genuine numbers. They yield a definite value for the size of the ultra-Higgs background. From this follow definite predictions for the mass of the new X gluons and for the lifetime of the proton.

What comes out is that the new gluon is about 10^{15} times as heavy as a proton. To give a sense of scale: this single elementary particle weighs about as much as a living amoeba. The latest accelerators have only recently attained the energies to produce a few W bosons—and the X bosons are 10^{13} times as heavy, far out of reach. The best hope for checking whether the new gluons exist is to search for the rare interactions they induce during their occasional, fleeting existence as virtual particles.

With that thought, we circle back to proton decay. The lifetime of the proton, limited by exchange of the new gluons, is predicted to be about 10^{31} years. This corresponds to about ten decays per year in each one thousand tons of matter.

If a proton decays, it will generally emit products—positrons, for instance—moving at close to the speed of light in vacuum. The speed of light traveling through water is, however, considerably less than its speed in vacuum, and so the rapidly moving positron outraces light. Just as a jet plane flying faster than the speed of sound emits a sonic boom, the positron moving faster than the speed of light sends out a photic boom—a shock wave in light. And it is such shock waves, originating anywhere in the subterranean vat of water they guard, that huge sensitive photomultipliers are waiting for.

It is an unlikely device, in an outlandish place. The technicians who handle repairs and maintenance in the tank must double as skin divers. And yet it is one of the noblest monuments to man's curiosity and ingenuity. If our civilization disappears, let us hope for its reputation's sake that a later generation of archaeologists uncovers this bizarre swimming pool (rather than, say, the computerized death machines hunkered down beneath Cheyenne Mountain).

Unfortunately, several years of watching have yielded not a single observation of proton decay. The lifetime of the proton is now believed to be greater than 10^{31} years. Since it is difficult to make detectors much bigger, or to watch much longer, the prospects for direct observation of proton decay no longer appear very bright.

Although this is a terrible disappointment—as if we traveled with high hopes to a distant theater, only to find that the play had been canceled—physicists will not easily give up on grand unified theories. It is painful to think of losing the natural beauty of the Unification Table, or the explanation of the hierarchy of couplings that grand unified theories provide. But is it possible to retain these successes, in view of the stubborn reluctance of protons to decay?

The lifetime predicted for the proton, you will recall, was based on an estimate of the rate at which the couplings for the strong, electromagnetic, and weak interactions converge. These couplings approach one another at short distances because of the shielding or antishielding effects of virtual particle, which tend to make the weak coupling grow and the strong coupling shrink. Now, of course, the size of these effects depends on precisely what kinds of virtual particles there are.

In our previous calculations, we assumed—in good radically conservative style—that the particles we know about or (in the cases of the Higgs particle and the top quark) have excellent reasons to suspect exist are all the particles we need to consider. By relaxing this assumption, we can accommodate longer proton lifetimes without doing violence to the basic ideas and successes of grand unification.

For the moment, then, the noble attempt to explore the implications of grand unified theories by direct experiments has brought us only to an anticlimax. Nevertheless, these theories undeniably give wonderful insights—summarized in the Unification Table and in figure 25.2—into the observed pattern of particles and interactions, and they remain extraordinarily attractive.

The new theories offer another avenue for testing and exploration, based on their premise that physics changes markedly at extraordinarily high energies. Such energies are beyond the reach of accelerators but not beyond that of the big bang. During the early moments of the big bang, it is possible—more, it is the radically conservative view—that what we currently regard as extraordinary energies were commonplace. Grand unified theories were in their true glory. The ultra-Higgs background melted away—or, rather, it hadn't yet come to exist (the tenses get confusing, when you're postdicting)—and true democracy among all colours was restored. The interactions responsible for proton decay, too tiny for practical observations today, proceeded at a furious rate.

Does any relic from this brief flowering of unification survive to the present day? In the following theme, we will argue that the uniformity of the universe may be just such a relic. And in the next chapter, we will argue that the very existence of the matter you see around you—and indeed of the matter you are—is possibly another.

27

A Little Matter

We made a start on Leibniz's great problem—why there is something rather than nothing—in chapter 18. There we saw that particle-antiparticle pairs could be produced from pure energy. This suggested the beginning of an approach to Leibniz's question: we can start with pure energy and evolve material from it.

But if the existence of antimatter opens an approach to the origin of the material universe, it at the same time poses cosmological riddles. First of all, what happened to the antimatter? Why is there so little of it in the natural world? An explanation of the dominance of matter over antimatter must be part of a satisfactory answer to Leibniz, for if matter and antimatter coexist, they make an unstable mixture. They tend to annihilate one another, converting material back into exotic forms of energy.

The dominance of matter over antimatter is especially puzzling because it violates a fundamental symmetry in the laws of physics. There is a symmetry of physical laws, believed to be exact, that interchanges particles and antiparticles. (We refer to CPT symmetry, discussed in chapter 18.) But from the point of view of this symmetry, the content of the universe is completely lopsided. You certainly do <u>not</u> reproduce the past of our universe by changing particles into antiparticles as you reverse the arrow of time; our universe hasn't contained many antiprotons for quite a while. So one riddle leads to another: Where has the symmetry between matter and antimatter gone—or was asymmetry perhaps built into the universe from the start?

But wait a minute—before we get too agitated by these riddles, let's examine their premise. Are we really sure that the universe consists entirely of matter? If the matter immediately around us were matched by antimatter far away, our riddles would dissolve.

It is easy to demonstrate that matter and antimatter can't be in intimate contact anywhere. When a particle meets its corresponding antiparticle, they tend to annihilate each other and convert their mass into energy. For instance, a star made up half of matter and half of antimatter would immediately disappear in a titanic explosion, equal to thousands of supernovae. So such a star would exist for only an instant, and it would be very conspicuous. Neither such an explosion nor the lingering glow of the ashes it would leave behind has ever been observed.

The possibility remains, however, that matter and antimatter might coexist in the universe if each were confined to isolated regions, separated by empty space.

In the search for signs of distant antimatter, a good start is to examine cosmic rays, high-energy particles that arrive from space. Although the origin of cosmic rays is not yet fully understood, they certainly come from throughout the galaxy, and some of them probably have a still more distant origin. Cosmic rays are overwhelmingly made up of particles such as protons and electrons, or atomic nuclei made up of protons and neutrons. Antiprotons, and especially antinuclei, are almost never encountered. The stray antiprotons that do occur in cosmic rays most likely are secondary products of high-energy collisions. In other words, a few antiprotons are produced in the same way that antimatter is produced at accelerators, but there is little sign in cosmic rays of "primordial" cosmic antimatter. It therefore seems firmly established that our Milky Way consists entirely of matter; and it is only a little less certain that the Local Group of galaxies, of which the Milky Way is a member, is also all matter.

It is much harder to tell whether more distant galaxies are composed of matter. "Looking at" a galaxy, using the conventional tools of astronomy, we get no hint of whether it is made up of matter or antimatter. This is because virtually all such tools boil down to the detection of electromagnetic radiation—photons. The photons involved might be in many forms: the old mainstay visible light, or radio waves, microwaves, infrared or ultraviolet radiation, X rays, gamma rays—whatever. The problem is that there is no way to distinguish a photon emitted by matter from one emitted by antimatter. The laws governing the emissions and interaction of photons are (to very high accuracy) symmetric between matter and antimatter. As a result, the light from an antimatter galaxy would be identical with that from a matter galaxy, in every detail across the entire spectrum. For example, the characteristic emission lines of the hydrogen atom (proton plus electron) are duplicated exactly by the emission lines of the antihydrogen atom (antiproton plus positron). So "looking at" distant galaxies does not tell us whether they are matter or antimatter.

On the other hand, if an antimatter galaxy were close to a matter galaxy, the boundary region between them would be the site of frequent particle-antiparticle annihilations. The border region would therefore be an intense source of high-energy photons—gamma rays. Many astronomical sources of gamma rays are known, but none has the characteristics we would expect from a matter-antimatter border. It seems established that clusters of galaxies must consist entirely of matter or entirely of antimatter, and not of a mixture of the two.

This sort of argument fails, however, if clumps of matter and antimatter are separated by large regions of empty space. And, to be honest, it is difficult to exclude by any sort of observational evidence the idea that the universe consists of huge islands of matter and antimatter—each larger than a cluster of galaxies—separated by spaces so empty that tell-tale annihilations rarely occur. Nevertheless, the overwhelming majority of physicists and astronomers have come to believe that matter dominates over antimatter everywhere. The most important reason for this opinion is, ultimately, historical. It is hard to imagine how matter and antimatter, existing together in the early stages of the big bang, could ever have segregated themselves into distinct, well-separated regions. Such segregation would be difficult to reconcile with the uniformity of the universe, particularly the spectacular uniformity of the microwave background. The straightforward, radically conservative picture seems much more likely: matter and antimatter simply annihilated each other everywhere, until the antimatter was completely consumed.

It is clear enough what the antiparticle of any particular particle is. For instance, the opposite of ordinary matter is stuff made of antiprotons, antineutrons, and positrons. But we will find it very helpful to have a more general, abstract definition of matter and antimatter, which will apply to the entire zoo of particles that existed in the early universe. *Quark number* is the appropriate concept. The quark number of a collection of particles is defined to be the number of quarks the particles contain, minus the number of antiquarks. (Beware of this slightly tricky use of language: the quark number is not necessarily the same as the number of quarks.) For instance, a proton has quark number 3, as does a neutron, while an antiproton has quark number -3 and an electron has quark number 0.

The dominance of matter over antimatter can be restated in terms of quark number. It becomes the statement that the quark number of the universe is positive or, in other words, that there are more quarks than antiquarks in the universe. This will mean that there are more protons and neutrons than antiprotons and antineutrons—and will imply that ultimately, after annihilations do their work, just the protons and neutrons survive.

Neither the movement of quarks and antiquarks nor even their creation and annihilation in matched pairs changes the quark number. In fact, you can quickly check that none of the fundamental processes we have found to govern the strong, electromagnetic, and weak interactions changes quark number. Nor does gravity. In all these interactions, quark number is conserved. In any reaction due to any combination of these interactions, the total quark number of what comes out will be exactly the same as the total quark number of what went in.

Is the law of conservation of quark number ever violated? No one has ever found an exception—and not for lack of trying. As we discussed before, people have gone to extraordinary lengths in their attempts to detect proton decays. They have even been known to go into voluntary exile in salt mines near Cleveland. Because all the particles lighter than the proton have quark number 0, any decay of the proton would necessarily involve a change in quark number. The remarkable, and extraordinarily frustrating, stability of the proton is powerful evidence that interactions that violate quark number are quite rare.

If quark number is exactly conserved, the quark number of the universe was built in from the beginning, with no possibility of evolutionary change. If that's the case, we must abandon hope of reaching any historically based understanding of the cosmic asymmetry between matter and antimatter.

Fortunately for the possibility of understanding, grand unified theories of particle interactions, whose great appeal we have already discussed in other connections, give us compelling hints that quark number is <u>not</u> exactly conserved. These theories contain ultraheavy particles, the X bosons, whose exchange allows protons to decay. The interactions of X bosons do not conserve quark number. Because X bosons are ultraheavy, their virtual exchange leads to ultraweak interactions and ultrarare processes. Virtual X bosons represent enormous quantum fluctuations in the energy of the X field, and such fluctuations are rare. They'd better be, because we know from experiments that the lifetime of the proton is exceedingly long, more than 10^{31} years.

Although interactions that change quark number are ultrarare now, they could have been much more common in the early universe. This is not because the laws of physics were any different, but rather because in the very early, very hot stages of the big bang plenty of energy was available. Even particles as heavy as the X bosons of unified theories would be abundant at 10^{28} degrees Kelvin. At such temperatures, the energy barrier, which today makes changes in quark number rare, was easily breached.

The picture suggested by grand unified theories, then, is that processes violating quark number are common in the earliest moments following the big bang but become exceedingly rare as the temperature drops. This picture brings the problem of matter-antimatter asymmetry into sharp focus.

First, it makes it all but unavoidable that the universe <u>started out symmetric</u> between matter and antimatter, for when quark number is rapidly fluctuating, in thermal equilibrium it averages to zero. Initial symmetry is therefore not only an aesthetic aspiration but also a physical necessity.

Second, it makes it clear that the asymmetry between matter and antimatter is a <u>relic</u> surviving unchanged from the earliest moments of the big bang. After a brief moment of rapid flux, the quark number of the universe ceases to change. It is "frozen in."

These ideas should have a familiar ring—we've heard this song before. Very similar ideas arose in our discussion of the origin of universal chemistry. In each case, we describe the passage between different levels of equilibrium. At high temperatures, everything is fixed statistically; at low temperatures, everything is frozen in. The passage between these levels is determined by the laws of physics operating in an expanding and cooling universe. There is no room for ambiguity.

By reformulating the matter-antimatter asymmetry in terms of quark number, we arrive at a startling new perspective on the content of the universe. We come to realize that the survival of matter in the universe is actually, in a precise sense, a tiny effect. Remember that quark number, the quantity that is a relic of the earliest moments of the big bang, is the difference between the number of quarks and the number of antiquarks. At the high temperatures prevailing shortly after quark number were first frozen in, there were both lots of quark and lots of antiquarks. In fact, working backward, one can calculate that there were roughly a billion quarks and antiquarks for each present unit of quark number. In other words, today's asymmetry between matter and antimatter arose from a tiny mismatch early in the history of the universe, when there were a billion and one quark for every billion antiquarks.

Even once we accept that the cosmic asymmetry between matter and antimatter is, despite appearances, a tiny effect—just a little matter—we must still understand how it is possible at all. How can symmetrical laws, working from a symmetrical start, produce an asymmetrical result?

The most fundamental symmetry between matter and antimatter is CPT symmetry. CPT symmetry says that you generate a world interchangeable with ours if you simultaneously change particles into antiparticles (charge conjugation C), reflect right into left (parity P), and reverse the direction of time (time inversion T). As we discussed before, CPT symmetry is closely bound to the symmetry of space and time in relativity theory. For this reason, CPT is believed to be an exact symmetry of physical law. It might seem, then, that any attempt to favor matter over antimatter is doomed—there would always be an equally likely *looking-glass world,* generated from ours by the CPT operation, where antimatter dominated.

The loophole in this argument is that although the laws of physics are the same in the looking-glass world, the history of the universe is not. If we reverse the direction of time, we totally change the nature of the early universe. Instead of expanding, it contracts. So it is perfectly consistent with CPT symmetry that matter will come to dominate in an expanding universe. CPT symmetry then informs us that antimatter would come to dominate a contracting universe, but this symmetry does no more than relate the two possibilities. The cosmic asymmetry between matter and antimatter is a consequence of the cosmic asymmetry between past and future.

Having disposed of the most fundamental symmetry relating matter to antimatter, we must still check that no other symmetry exists that would rigorously forbid even the tiny effect we seek.

The history of the idea that physical laws are indifferent to the distinction between matter and antimatter is a history of upset expectations. Until the mid-1950s, it was thought that the laws of physics took no account of this distinction—in other words, that the operation C of changing particles to antiparticles was a symmetry of physical law. If this were true, it would, of course, be impossible for our cosmic asymmetry to develop. Careful study of the details of the weak interactions, however, revealed that the symmetry of the operation C is not an exact symmetry of physical laws.

An example of its violation is in the decay of muons and antimuons. (See figure 27.1. We shall have much more to say about muons in chapter 30; for now, you need to know only that a muon is basically a heavier version of the electron.) Muons spontaneously decay into electrons and invisible stuff—neutrinos of various sorts—with a half-life of about a microsecond. Antimuons decay into the corresponding antiparticles, with the same lifetime. So far, everything seems perfectly symmetrical. When the electrons emerging from muon decay are carefully studied, however, it is found that they are almost always left-handed, whereas the positrons emerging from antimuon decay are almost always right-handed. So there is a definite distinction between the two cases, and C is not a symmetry of physical laws.

In response to this surprise, symmetry-loving theorists were quick to propose a modified symmetry that seemed to be respected by all interactions. The idea was that the laws of physics would be unchanged if you simultaneously changed particles into antiparticles and reflected left into right—the CP operation. In other words, it was proposed that for any physical process involving particles, the mirror-image process involving antiparticles would also be possible. More precisely, in quantum mechanics, it was proposed that the probability for these two processes should be the same. For example,

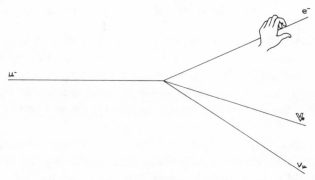

Figure 27.1

the decays of muons we just discussed respect CP even though they violate C. Both muon decay and the antimuon decay you get upon simultaneously changing particles into antiparticles and right into left proceed at the same rate. If CP symmetry were exactly valid, a preponderance of matter could not evolve from primordial equality. For every process that created a particle, an equally likely mirror-image process would create the antiparticle.

The concept of exact CP symmetry survived for about seven years. Then it was observed that the long-lived neutral K meson, which is its own antiparticle, decays more often—by a fraction of a percent—into a negative pion, a positron, and a neutrino than into a positive pion, an electron, and an antineutrino. If CP were an exact symmetry, the two decay modes would have to be equally likely. Although to this day the only processes observed to violate CP invariance—slightly—are obscure decays of rather exotic particles (K mesons), it is reassuring that this barrier to the creation of matter is not completely insurmountable. After all, we need only a tiny effect.

So far, we have discussed several general principles likely to be relevant to any scientific theory of how the cosmic asymmetry between matter and antimatter arose. Have we captured all the essentials? If we were engineers considering the design of an airplane, we might at this point build a prototype; if we were mathematicians, we would try to construct an existence proof. Being at the moment physical cosmologists, we imagine a universe. Our goal is not to make all the details realistic—an all but impossible task, given our limited knowledge of physical laws in the relevant ultra-ultraextreme conditions—but to check that there are no hidden pitfalls.

A very simple scenario for how the cosmic asymmetry between matter and antimatter could have developed is called the "drift and decay" scenario. Inspired by grand unified theories, we suppose that there is a very heavy particle—call it X—whose interactions violate the conservation of quark number. To be specific, let's suppose that X sometimes decays into two quarks and sometimes into an antiquark and a positron. (That's what the extra colour-changing gluons in grand unified theories are predicted to do.) Since X can decay alternatively into products with quark number 2 or -1, it is obviously impossible to have quark number conserved in both sorts of decays. So quark number is not conserved; it can change.

When the universe is at an extremely high temperature—say, again inspired by unified theories, 10^{28} degrees Kelvin—there's plenty of energy available to create X, and also anti-X or **X** particles. They would be in thermal equilibrium, and exist in equal numbers. As the universe cools, it rapidly becomes more and more difficult to create X and **X**, until further production ceases about 10^{-33} seconds after the big bang. The ones created will either annihilate one another or simply drift until they spontaneously decay.

Let's concentrate on the decays, since the annihilations are slightly more complex (and don't really change the picture). Suppose, to be definite, that the two kinds of X decays we discussed above are equally likely. Then on the average each X decay will contribute 1/2 unit to the quark number of the universe, as you can see after doing some simple arithmetic. However, this does not yet solve the problem of the origin of quark number in the universe, because we have to worry about what **X** is doing. You see, for each possible mode of X decay there is a corresponding mode of **X** decay. If X can decay into two quarks, **X** can decay into two antiquarks, and so forth. If the probabilities for these corresponding decays were the same, **X** decays would neatly cancel the quark number produced in X decays. Are the probabilities in fact the same? They would be, if the laws of nature were symmetric under C, the interchange of matter and antimatter. They would be even if the laws of nature were symmetric only under CP, because just having particles fly to the left and equal numbers of antiparticles fly to the right doesn't generate any quark number either. But we have seen that C and CP are not exact symmetries of nature. There are tiny interactions, mysterious in detail, that violate these symmetries. A tiny effect is what we're looking for. If the probabilities that **X** will decay into two antiquarks or a quark and a positron, instead of being 50–50, are respectively 49.9999999 percent and 50.0000001 percent, we will get a density of quark number—a preponderance of quarks over antiquarks—that's just about right. Enough quarks will then survive the later wholesale annihilations to account for the observed amount of matter in the universe. It's a remarkable thought that the material world we see around us is the residue of this difference in the ninth decimal place.

The simple drift-and-decay scenario demonstrates that once we have satisfied the basic requirements—there must be interactions that violate both conservation of quark number and CP symmetry—an asymmetry between matter and antimatter can easily arise, starting from an initially perfectly symmetric universe.

The arrow of time plays a subtle, but crucial, role in this scenario. It was only because the universe was expanding and cooling that the back processes—the creation of X and $\overline{X}$ from quark collisions, the inverse of the decays—could be neglected. Otherwise, these back processes would destroy quark number at exactly the same rate as the decays create it.

The existence of matter in the universe today has now been traced to a tiny imbalance, a minuscule mismatch in the early universe. The cosmic asymmetry between matter and antimatter, according to these ideas, arises from a more fundamental asymmetry—the cosmic distinction between past and future. Physical laws, operating on an expanding and cooling but otherwise perfectly symmetric universe, inevitably produce an asymmetry between matter and antimatter.

A further question remains. We have described how the universe could have begun with symmetry between matter and antimatter and then have grown asymmetric. But why should the universe have been symmetric in the beginning?

At one level, this question is answered statistically, by the idea of thermal equilibrium. At a deeper level, however, this explanation is not fully satisfying. It fails to explain why the universe is symmetric in several other ways: it is electrically neutral, on the average, and it seems to have no net rotation. It also fails to explain why an explosive event, the necessary source of energy to start the universe, should have occurred at all. The big bang scenario has been justified by observations—the universe is still expanding—but has never been shown to be a <u>logical</u> necessity.

We shall now describe an idea that may lead to an understanding of these questions. It is by no means well established, but it can, in the spirit of radical conservatism, inspire a program of research. It was the original motivation for my (Frank's) own work on the origin of matter-antimatter asymmetry.

Modern theories of the interactions among elementary particles suggest that the universe can exist in different phases, analogous to the magnetized and unmagnetized forms of iron oxide, distinguished from one another by the presence or absence of Higgs and ultra-Higgs fields. The laws of physics are more symmetric in some phases than in others. Background fields tend to spoil symmetries, just as the field of a magnet distinguishes a direction in space.

In these theories, the most symmetric phase of the universe generally turns out to be unstable. One can speculate that the universe began in the most symmetric state possible and that in such a state no matter existed: the universe was a very empty vacuum, devoid both of particles and of background fields. A second state of lower energy is available, however, in which background fields permeate space. Eventually, a patch of the less symmetric phase will appear—arising, if for no other reason, as a quantum fluctuation—and, driven by the favorable energetics, start to grow. The energy released by the transition finds form in the creation of particles. This event might be identified with the big bang.

The electric neutrality of the universe of particles would then be guaranteed, because the universe lacking matter had been electrically neutral, and electric charge is exactly conserved. The lack of rotation in the universe of matter could be understood as being among the conditions most favorable for the appearance and growth of the fields that herald the new phase.

Our answer to Leibniz's great question "Why is there something rather than nothing?" then becomes "'Nothing' is unstable."

Radical Uniformity in Macrocosm

As time goes on, ever more distant regions of the universe come within our purview, regions from which we receive light as old as the universe itself. And yet the new always seems to reproduce the old. The striking uniformity of the universe, that extends even to regions between which there has been no means of contact, presents an enormous riddle.

Recently, a promising answer to this riddle has been proposed. It is the idea that the observable universe was once a much cozier place, that after being in intimate contact its contents were dispersed in a period of cosmic *inflation*. The inflationary-universe idea explains the uniformity of the universe as being akin to the smoothness of an inflated balloon, which results no matter how wrinkled the limp rubber was to begin with. And it suggests that deviations from perfect uniformity, which provide the seeds of galaxies, arise only from the unavoidable uncertainty of quantal reality: a remarkable connection between macrocosmos and microcosmos.

GENESIS MACHINES

The plot of *Star Trek II: The Wrath of Khan* pivots on a device called the genesis machine, which can in a short time rearrange the structure of matter over entire planets—making them fertile and suitable for habitation but, of course, destroying any preexisting life in the process. The principles behind the operation of the genesis machine are not explained in the movie. How could something along these lines work?

In fact, a genesis machine of sorts has already acted on planet Earth. The original atmosphere of Earth was deficient in oxygen but rich in such noxious gases as the malodorous hydrogen sulfide. Such an atmosphere would be extremely poisonous to most present-day life. It was transformed into its current form by the metabolism of generations of simple but robust and rapidly reproducing life forms—first by anaerobes, which can flourish only in oxygen-free environments (a surviving representative of this once dominant form of life is the bacillus responsible for botulism, which thrives in airtight jars or cans), then by simple, one-celled plants like blue-green algae. The anaerobes absorbed the noxious gases and combined them into larger, less reactive molecules. As a by-product of photosynthesis, the simple plants liberated oxygen, which happened—accidentally, from the point of view of the plants—to be a potent source of energy, once evolution developed the trick of respiration.

The natural transformation of Earth's atmosphere probably took place over an interval of several hundred million years. A genesis machine acting this slowly would be of limited practical interest.

Given a suitably designed microbe, however, there seems to be no fundamental obstacle to making such a process run much faster. The key that opens up such possibilities is the process of runaway, exponential growth. A small number of microbes, provided with a favorable environment in which to carry on their multiplication through division, can quickly become a multitude. With the first cycle of reproduction, one becomes two; then these two become four, four become eight, eight sixteen. From this modest start, one has after thirty cycles become a billion; after forty cycles, a trillion.

Numbers capable of transforming entire planets are produced in short order. Imagine a microbe capable of processing a few times its own weight of atmospheric material, and of reproducing by division, in a day. Within about seventy generations—a few months' time—a single initial gram of such bacteria could transform the atmosphere of an entire planet.

There are "genesis machines," working on the same underlying principle of exponential growth, that transform matter not just by juggling intact atoms from one molecule to another but in even more fundamental ways. We are thinking of nuclear reactors and fission bombs. The role of the "microbes" in these devices is played by neutrons. In a suitable "atmosphere"—what we call a fissile material—neutrons are likely to collide with a nucleus, which breaks up into products including several more neutrons, liberating energy in the process. The resulting second generation of neutrons goes on to produce more breakups and still more neutrons, in a potentially runaway situation. Such an exploding population of neutrons can detonate the material in a fission bomb within a small fraction of a second. Although such a release of energy is swift and sure, it is not ideally suited to the generation of electric power. Nuclear reactors use moderating rods, containing nuclei that can absorb excess neutrons, in order to control the neutron population explosion.

Could there be a genesis machine in the field of thought? Suppose, as seems inevitable, that we learn to design computers cleverer than we are. Presumably, they will be able to design even cleverer computers, which in turn will design machines still more clever. . . . Is this a runaway process, or does it become more and more difficult for higher levels of intelligence to transcend themselves? And what form would the later generations take?

Simple facts of human evolution suggest speculations along these lines. The development of human intelligence, from comparatively primitive, apish beginnings, was remarkably rapid on evolutionary time scales. It can therefore have involved only small alterations in the genetic code, in particular in the instructions for assembling a brain. Indeed, the overwhelming bulk of the genetic message in the DNA of men and apes is held in common. For example, the code for the much-studied protein cytochrome-c, 104 amino acids long, is identical for human and chimpanzee.

How can a small change in the instructions give such a dramatic change in the result? It seems most unlikely that the difference can be attributed to a small difference in the structure of a few proteins, which would alter their specific chemical activity. The changes in brain architecture and information processing between man and ape go deeper than chemistry. A much more reasonable possibility is that the critical changes are changes in genes that exercise <u>control</u> in the process of fetal development, genes that regulate the expression of the more conventional genes, much as a computer program controls the expression of the computer's fixed hardware. Many such control or regulator genes are known in simpler contexts.

Suppose there are processes that produce stereotyped working modules of brain cells, and genes that control the switching on and off of such processes. Quite simple alterations of such genes could drastically increase the number of modules that get produced and, thereby, the power of the resulting brain.

(There is good evidence, for both men and apes, that large parts of the cerebral cortex and the cerebellum are in fact organized into many semiautonomous modules, arranged in geometrically regular patterns. This organization has been elucidated most thoroughly for the part of the brain that receives visual input. For example, it's known that the visual cortex is organized into several thousand semiautonomous columns, about a millimeter on a side at the surface of the brain. Roughly speaking, each module is primarily concerned with the processing of features in the image received by one eye that arrive at a specific location on the retina.)

If something like this is the mechanism whereby human intelligence arose, at least three important conclusions follow.

First, there must be an underlying principle of brain organization that makes it possible to use large numbers of essentially identical modules effectively. To the extent that a single human being is smarter than a band of chimpanzees, nature must have found some scheme for wiring the modules together so that they can function smoothly and cooperatively. If we could begin to understand what these principles are, we would have gone a long way toward explaining the nature of human intelligence and, presumably, have learned something very useful for designing truly intelligent computers.

Second, human intelligence is fundamentally limited by human brain size, which is in turn limited by the size of the human female pelvis. Indeed, if there were room for more modules, they could be used (according to our speculation) to increase intelligence. That this limit is being severely pushed is evident from the difficulty of labor and from the prolonged immaturity and helplessness of the newborn. There is clearly a high evolutionary premium on having the largest-possible brain; this would be difficult to understand if it were something other than sheer number of units that was the crucial limitation on intelligence. All this suggests that the great Brains, the Fourth Men imagined by Olaf Stapledon, might not be so fanciful after all:

> Those who sought to produce a super-brain embarked upon a great enterprise of research and experiment in a remote corner of the planet. . . . [They] employed four methods, namely selective breeding, manipulation of the hereditary factors in germ cells (cultivated in the laboratory), manipulation of the fertilized ovum (cultivated also in the laboratory), and manipulation of the growing body . . . . By inhibiting the growth of the embryo's body, and the lower organs of the brain itself, and at the same time greatly stimulating the growth of the cerebral hemispheres, the dauntless experimenters succeeded at last in creating an organism which consisted of a brain twelve feet across, and a body most of which was reduced to a mere vestige. . . .

To create them might require only (!) that birth be liberated from the human pelvis and that the relevant control genes be switched on longer.

Third, it should be possible to build a genesis machine in the field of thought, based on the same principle of exponential growth as are the other genesis machines. This requires that the modules contain instructions for their own duplication—be self-reproducing—in order that their number could increase at an exponential rate. A wild vision, worthy of Stapledon, in which entire planets are transformed into brains of incalculable power, begins to take form.

It remains to discuss what is perhaps the ultimate genesis machine. Let us begin with a well-understood and perhaps familiar example. Water vapor can remain in the gaseous phase even at temperatures below the boiling point. At such temperatures, the liquid phase is stable, but the gaseous phase can exist for a very long time if it is pure and carefully handled. In other words, the vapor of the gaseous phase is metastable. Water vapor in this condition is said to be *supercooled.*

Supercooling can occur because the liquid phase is stable only once it attains a certain minimum size. Droplets of liquid that are too small quickly evaporate back into the vapor. For the supercooled vapor to turn liquid, the crucial step is to form some initial seed droplet that is not too small; this droplet, once formed, will quickly grow to incorporate all the available water molecules. In practice, seed fluctuations grow around little impurities, such as dust grains or the rough surface of containing walls, or are formed when the vapor is jostled. The "seeding" of clouds consists of dosing them with suitable seed impurities, in the hope that the vapor will condense around these impurities—forming rain drops—rather than remain a supercooled cloud.

Now, imagine life developing in an environment of supercooled vapor. Suppose that intelligent life forms eventually appear, begin to study physics, and come to understand about supercooling. They then realize that the existence of their society—in fact, of the world as they know it—is extremely precarious. In the long run, they cannot prevent the ultimate ecological catastrophe. Eventually, a sufficiently large droplet will appear and grow to engulf their world.

A related sort of ecological catastrophe has happened before, on a universal scale, in the course of cosmic evolution. The role of the water molecules is played by certain unfamiliar particles known as Higgs particles, but the physics is otherwise very similar. The Higgs particles, in some circumstances, will condense only around sufficiently large droplets. When they do, the properties of everything change, because everything interacts with the ambient Higgs particles. Elementary particles that used to be massless become massive, or vice versa. The boundary of the condensing drop expands, essentially at the speed of light, and marks the division between two worlds in which the basic constituents of matter have very different properties.

As we said, this sort of natural genesis machine has acted before, probably several times, in the history of the universe. Can it happen again? Are we, like the poor supercooled vapor creatures, faced with inexorable doom? Our current knowledge of the laws of physics is insufficient to tell us for sure.

If the ultimate ecological catastrophe does occur, we'll receive no warning. No message can outstrip the condensation front, which moves at the speed of light. And, of course, once the condensation front arrives, the properties of the material in our bodies is so drastically altered that consciousness is inevitably destroyed. We'll never know what hit us. In a peculiar way, then, there's nothing to worry about.

28

Horizons

God hides within night's blackest folds
Most wisely, what the future holds.
—Horace, *Odes* 3.29 (23 B.C.)

In a universe of finite age, where information travels no faster than the speed of light, we can learn about only a limited volume of space. Light from galaxies too far away will not have had enough time to reach us in the entire history in the universe. There is, in other words, a cosmic *horizon* that marks the limit of the accessible universe. We have no way of observing what's going on in regions of the universe beyond our horizon.

A little tragedy may make the idea clear. Imagine two people, call them Dante and Beatrice, both born immediately after the big bang in different parts of the universe. Dante is a cosmologist. In his observations of the distant universe, Dante—then aged thirty-five— one day witnesses the birth of Beatrice. He is so taken with the beauty of this baby that he falls hopelessly in love with her. In those days, lives were short, and, sad to say, Beatrice never lived to be thirty-five. She might have fallen in love with Dante, had she had a chance to see him. But she never got the chance. Just as her galaxy first came into Dante's view thirty-five years after the big bang, observers in her galaxy could first view Dante's galaxy only thirty- five years after the big bang. And so, while Beatrice lived, Dante was *outside her horizon.*

Here is another way of thinking about cosmic horizons. You are probably familiar with the idea of measuring the distance to astro- nomical objects in light-years; we say an object is so many light-years away if light from it takes that many years to reach us. When the distance gets to be ten billion light-years, we are getting close to the horizon, since the age of the universe is about ten billion years.

There is another effect at work here, which stretches our horizons just a bit more. Because of the expansion of the universe, we can see things even farther away than the age of the universe in light-years. By the time we receive ten-billion-year-old light, its source has retreated farther still, and is by now more than ten billion light-years away.

As the universe expands, it is expanding our horizons along with it. With the passage of time, regions ever more light-years away become accessible. In other words, looking up at today's sky, we can in principle see things that were beyond yesterday's horizon.

Figure 28.1 illustrates the way our horizons grow. The solid lines represent light rays diverging from a point at the bottom of the picture, which represents the time of the big bang. Two Earth-bound observers, astronomer P and her far-distant descendant astronomer Q, observe this big bang light at very different times.

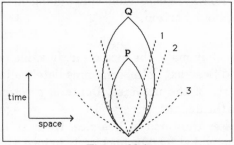

Figure 28.1

The light rays are bent by the expansion of the universe, and enclose a large but finite volume of space. To understand their curvature, it is helpful once again to use the trick of imaginatively running time backward and to think about a light source on Earth sending light back in time. At first, the beams diverge from their source, but they are inevitably brought back together by the *shrinking* of the universe.

The dashed lines represent world lines of distant galaxies (or more generally, the sort of path followed by material moving along with the cosmic fluid—the stuff that fills the universe.) Only those galaxies whose paths intersect the smaller teardrop can have sent light signals to P during the history of the universe; only they are within P's horizon. In the figure, galaxy 1 is within both horizons, galaxy 2 is within P's but outside Q's, and galaxy 3 is outside both.

The growth of horizons makes the uniformity of the large-scale structure of the universe—the smoothness of the microwave background and the even distribution of galaxies—quite a puzzle. Ever-new regions of the universe have been coming into view over the course of cosmic evolution, and yet they always seem to resemble one another closely. Furthermore, all of them strongly resemble what we think our immediate neighborhood looked like in the distant past. It's like a farfetched science fiction story, where people land on unknown planets in distant galaxies and always find—other human beings!

This puzzle becomes even more perplexing if we consider how horizons limit the vision of observers elsewhere and elsewhen. We see distant galaxies as they were in their remote past. Because horizons have been expanding, the horizons of observers in those galaxies take in a much smaller volume of space than does our own. The interesting possibility arises that within our horizon we might see two galaxies in different parts of the sky whose own horizons contain nothing in common. If this occurs, we are seeing two galaxies that are truly "from different worlds"—no physical event in the history of the universe could have influenced both. Such galaxies are said to be *causally disconnected,* because the causes affecting their behavior are entirely distinct.

Are any examples of such causally disconnected regions—truly "different worlds"—available for our inspection? This weird theoretical possibility is in fact realized, and in a particularly unnerving way.

The photons in the microwave background carry information from early in the history of the universe, and therefore from very distant regions, beyond the reach of ordinary telescopes. Using standard assumptions about the big bang, we can calculate that regions in the sky more than about 10° apart were causally disconnected when the photons we now observe as the microwave background were emitted. In other words, nothing in the entire previous history of the universe could have affected both of two photons coming from such regions.

And yet, the overall intensity and spectral distribution of the microwave background are found to be uniform to within a part in ten thousand. How can utterly different causes lead to the exact same effect?

To get a perspective on this question, let's reconsider our science fiction analogue. Suppose that our distant descendants found themselves living out that farfetched yarn, where every inhabited planet turns out to be peopled by human beings. Could they frame a rational explanation?

One idea might be that evolution leads unswervingly to a single, inevitable result. This seems unlikely. Evolution has been powerfully shaped by random accidents—the mutation of particular genes (which can be traced to small changes in single molecules) and the vagaries of climate and geography. But even if evolution had taken the same course on all planets, humans would not be found everywhere, because different stars and planets have different ages. So even if the end results of evolution were always essentially the same, there should be planets inhabited only by one-celled animals, dinosaur planets, and planets where stages of life beyond the recognizably human have arisen.

A more reasonable explanation would be that long ago some kind of spores—genesis machines, perhaps, with instructions for directing the course of evolution toward humans—originated from a single source and were disseminated throughout the galaxy. The apparently bizarre coincidence that humans inhabit far-flung worlds would then be ascribed to a forgotten event in prehistory, which forged a hidden link among them.

With these problems and ideas in mind, let us return to the cosmological problem of horizons. One way to explain the similarities among unrelated parts of the universe is to claim that the course of cosmic evolution must be unique. One major theme of recent work in cosmology and particle physics is precisely to explain more and more of the structure of the universe in terms of the working of physical laws, removing the necessity or indeed the possibility of considering different "initial conditions." We have discussed how the abundance of the elements and the cosmic imbalance between matter and antimatter—features of our world that once were thought to be unchangeable and therefore unanalyzable—are now understood to be necessary consequences of the workings of physical laws in an ideally simple universal history. This line of thought, though much more reasonable than its biological analogue, falters for precisely the same reason. Even if causally disconnected parts of the universe are all constrained to follow a single evolutionary path, we are still left with a riddle: Why should they reach the same stage of evolution at the same time?

The uniformity of the microwave background radiation poses this problem in a very concrete form. If causally disconnected regions emitted their radiation at different times, the resulting photons would have gone through different amounts of expansion and cooling before reaching us. In that case, the microwave background would not be isotropic—the precise color of its pale fire would be different in different directions.

So we are left with the second idea, that a hidden prehistory links the different regions. Unfortunately, this contradicts our premise, that nothing lies within their common horizons.

The uniformity of the microwave background presents a paradox writ large across the sky. It is another example of the sort of paradox physicists love best: simplicity and symmetry where, it seemed, we had no right to expect them. Such paradoxes always focus our attention and force us into new ways of thinking. The *horizon problem* is no exception—read on.

29

Inflation

And if it has been blotted from thy view,
Now recollect . . .
. . . my oracles from hence
Shall be unveiled, far as to show their faces
May be appropriate to thy rude view.
—Beatrice to Dante, in *Purgatorio* XXXIII (ca. 1300)

We mentioned earlier that the distance to the horizon is actually larger than the age of the universe times the speed of light. Even though the speed of light sets an upper bound to the speed of all possible signals, we can receive signals from objects farther away, because in the past (when the light we now receive originated) these objects were closer. In fact, the expansion of the universe tends to stretch horizons along with it.

This effect is so central to our discussion of horizons and inflation that we'll give it a name: the *stretched-horizon effect*. The stretched-horizon effect emphasizes an important point about horizons: to know the distance to our horizon, we must know how fast the universe has expanded throughout its entire history.

Now, under ordinary conditions—using the term *ordinary* in a generous sense, to describe the expected behavior of matter at the extraordinarily high densities and pressures of the early big bang—the stretched-horizon effect is relatively small. The expansion of the universe is just not fast enough to magnify horizons much. More precisely, the periods of rapid expansion are very short-lived. In conventional big bang cosmology, the expansion of the universe is postdicted to be arbitrarily rapid at times close enough to the big bang itself, but the really rapid expansion doesn't last long enough to make a dent in the horizon problem.

These discouraging remarks certainly apply to the parts of universal history physicists claim to understand well (and can check)—that is, at least back to the first few minutes, when atomic nuclei were formed. They almost certainly apply back to the first few thousandths of a second. Enough is known about the behavior of matter at high energies that the course of cosmic evolution can be calculated with considerable confidence even that far back. And the calculations tell us that the horizon problem remains.

It is possible to imagine, however, that earlier still, when conditions were so extreme that no one really knows what was going on, there could have been a period when the universe expanded extremely rapidly. If the expansion was rapid enough, the size of the universe—and the size of horizons—could have been magnified by an enormous factor within a tiny fraction of a second. The word *inflation* is used to distinguish the extremely rapid expansion we are considering here from more routine rates of expansion that characterize the classic big bang. The idea that the universe underwent a period of inflation is called the *inflationary-universe scenario*. (Alan Guth first proposed this scenario in 1979, when monetary inflation was on everybody's mind, so the name was a natural.)

If the inflation idea is correct, every part of the universe now within our view arose from inflation of a region once so small that all parts of it were within one another's horizons. To put it another way, all the things we now observe were once close enough together to have been touched by common influences. And if this is the case, the horizon problem dissolves. We begin to have a chance of understanding the striking uniformity of the macrocosmos in general and the microwave background in particular.

Figure 29.1 shows how inflation stretches horizons. The teardrop of light that appeared in our discussion of horizons (figure 28.1) gets modified very close to the moment of the big bang, when inflation occurred. So do the paths of observers moving with the cosmic fluid. The most important thing to notice, in comparing these figures, is that in the new picture the dotted paths of galaxies that wind up at distances far beyond P's old horizon intersect the modified teardrop. Such galaxies used to be beyond P's horizon; now, thanks to events in the earliest moments of the history of the universe, they are within it.

Although what seemed to be a truly paradoxical observation of uniformity—what we have called the horizon problem—comes to seem

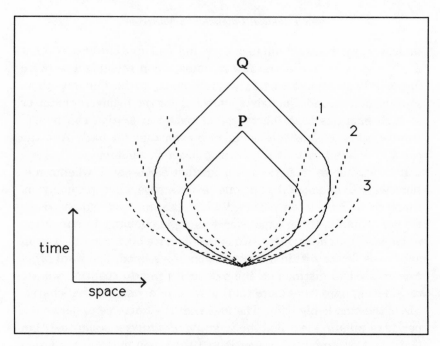

Figure 29.1

less paradoxical in the inflationary universe, a vestige of it survives. Horizons of influence are enormously widened in the inflationary scenario, because communications were established before inflation occurred, when the universe was much smaller. However, after the epoch of inflation, the newly distant regions cannot communicate any more. *Apparent horizons* mark the limits of communication after inflation.

In the figure, you'll notice that galaxy 2 can send a message to astronomer P only during the period of inflation. So, while galaxy 2 is inside the horizon of the astronomer at P, it is outside P's apparent horizon. It is not until much later that P's descendant Q will be able to see galaxy 2 as it exists after inflation. For astronomer Q, galaxy 2 is within both the apparent and the real horizon.

For instance, if inflation took place 10^{-34} seconds after the big bang—a popular, though astoundingly small, estimate of the time—two observers are said to be outside of each another's apparent horizons if no event after the first 10^{-34} seconds in the history of the universe could have affected them both. The name "apparent horizon" is quite appropriate. The apparent horizon is what we would think was the horizon, if we did not realize (or, rather, hopefully surmise) that inflation occurred. To see beyond the apparent horizon, we would need to see into preinflationary times—that is, to get a glimpse of the first 10^{-34} seconds following the big bang. You need to tell the modified teardrop from the original. It is difficult to do, because just about all traces of such very early times have been obliterated by subsequent events. In other words, it is hard to find relics of preinflationary times—and so, for most purposes, the apparent horizon really does mark the limit of knowledge.

If inflation did nothing more than dissolve the horizon problem, it would be an entrancing but farfetched speculation, more comforting myth than scientific theory. To go further, it is necessary to draw some testable predictions from the model and to see if it can be woven into the rich tapestry of better-established physical ideas.

Remarkably, there are other striking facts about the universe, readily explained by inflation, that would otherwise present vexing puzzles. One has to do with the average density of the universe. It turns out that the overall distribution of mass corresponds closely to the exact *critical density* that separates indefinite expansion from eventual recollapse. That is, if the universe were just a little denser than it is, gravitational attraction between different regions of the universe would not only slow down the expansion of the universe but eventually even reverse it. The fact that the density is poised near this unstable value is beautifully explained by inflation, as we shall discuss more fully in chapter 31.

Another theoretical advantage of inflation is that it explains why various "monstrosities" produced early in the big bang are not found in the universe today. The unified theories we found we found so attractive in chapter 25 tend to predict the existence of at least one breed of very heavy stable particles, called *magnetic monopoles*. We can't predict precisely how much such particles should weigh, but 10^{16} times the weight of a proton is a rough estimate. Magnetic monopoles get their name because they carry powerfully concentrated magnetic fields, which would make them readily detectable—if any were around to be detected. Because magnetic monopoles are so heavy, there is no prospect that any foreseeable accelerator will be able to create them. On the other hand, we can look around and see if we find them occurring naturally anywhere in the universe—

in cosmic rays, meteors, or moon rocks, for instance. Not one magnetic monopole has ever been detected. This fact sets stringent limits on their abundance in the universe. It is something of an embarrassment, because calculations show that the early moments of the big bang should have produced quite enough monopoles for them to be easily detectable today.

Just as we are about to be forced to choose between the allure of the unified theories and the elegant simplicity of big bang cosmology, inflation comes to our rescue. If an episode of inflation intervenes between the production of monopoles and the present, the original density of monopoles gets drastically reduced. Inflation, by diluting big bang monopoles to the point of undetectability, lets us keep our unified theories while doing minimal violence to big bang cosmology.

Inflation is a powerful idea for explaining the otherwise puzzling uniformity of the universe on large scales. By tracing the origin of the observable universe to expansion of an exceedingly tiny seed region, it explains why everything looks the same all over. Just as blowing up a balloon removes whatever wrinkles and crinkles it had before, so the process of cosmic inflation tends to stretch away any irregularities in the seed region.

We must make sure this explanation is not too good, however. The uniformity of the universe is not perfect. After all, when you look up at the night sky, you see not a faint, featureless haze but rather sparkles from a multitude of stars and galaxies embedded in vast regions devoid of light. And, of course, we and our immediate surroundings are much denser—by a factor of 10^{30}—than the cosmic average. Condensation of stars and galaxies, and ultimately of people, out of the tenuous cosmos can be triggered by the runaway growth of much less pronounced irregularities, as we discussed in chapter 6. But there have to be some irregularities, to get things started. Can inflation, having beautifully explained cosmic uniformity, also cope with its limitations?

At present, this question has no definitive answer. There is, however, a fundamental limit to the degree of uniformity that is attained after inflation. This limit arises because there are quantal fluctuations in the world of virtual particles on all size scales, down to the very smallest. In fact, such fluctuations—the fleeting coming to be and passing away of virtual particles—are most common and noticeable on small scales, as we saw in our discussion of antimatter. Now, during the process of inflation, quantum fluctuations in density on sub-sub-submicroscopic scales get blown up to cosmic proportions. It is entirely possible that such irregularities are the seeds from which galaxies will later grow. It may seem incredible that quantal effects, which we usually encounter only in the subtle interplay of sub-atomic particles, can underlie the growth of mighty galaxies—but calculations in the inflationary framework indicate it could well be so.

One slight problem remains. In the most straightforward models of how inflation works, which we shall discuss momentarily, microscopic quantal fluctuations make the model universe rougher than the real one. The cosmic irregularities they generate are more than sufficient to seed galaxy formation; indeed, they are so large that they cannot be reconciled with the observed uniformity of the microwave background. There are plenty of ways to tinker with the models, so this embarrassment is not necessarily a catastrophe. The idea that the fundamental limit to cosmic uniformity is nothing but the quantal structure of the microworld—that the universe is as uniform as it possibly could be—remains most intriguing and promising.

Clearly, the theory of inflation has much to recommend it. But why should inflation ever have occurred? Remarkably, there is a situation that might well have arisen in the early universe and that does trigger extremely rapid expansion of the universe.

Throughout most of the history of the universe, the bulk of its energy has been in the form of radiation or particles. These forms of energy exert positive pressure. In other words, as each part of the cosmic fluid tries to obey the command of inertia and expand, it is impeded by collisions with the surrounding matter.

Grand unified theories suggest another form that energy may once have taken. The ultra-Higgs fields now permeate all space, hiding some of the symmetry and simplicity of physical laws. This field configuration—a high constant value everywhere—represents their lowest-energy, and hence their most stable, state. At extremely high temperatures, however, the stability of low-energy states matters far less. Entropy wins out over energy, and the ultra-Higgs background melts. In the very hot early stages of the big bang, primordial energy was locked up in ultra-Higgs fields, energy that had to go somewhere before the fields could reach their present, low-energy state.

This form of energy differs from the energy associated with radiation or particles in one, crucial respect. Ultra-Higgs field energy exerts <u>negative pressure</u>. (I have tried and failed to come up with a simple qualitative explanation of why this should be. It is what comes out of the equations, and you will have to take my word for it—F.W.) In other words, this form of energy <u>encourages</u> rapid expansion of the universe. If ultra-Higgs field energy ever comes to be the dominant form of energy in the universe, a period of extremely rapid expansion—that is, inflation—will occur.

So little is known about unified theories that we can't postdict with confidence whether any ultra-Higgs field ever does dominate the energy of the early universe, or the details of how things would evolve if one did. Nevertheless, it is quite suggestive and encouraging that the fields we were led to consider for entirely different reasons, in trying to unify fundamental forces, could well give us inflation as a cosmological bonus.

Dear reader, you must have noticed that in the last few chapters conjecture has been piled atop conjecture, creating a mighty tower of speculation with few solid elements of support. Have we strayed from the narrow path of scientific investigation into the tempting pastures of pure mythmaking and wishful thinking? Perhaps. But we do this with open eyes.

Let us recognize that there is a <u>hierarchy of belief</u> among scientific theories.

At the top of this hierarchy are theories like QED and QCD, which are fundamentally simple and very definite and which explain a wealth of data in quantitative detail. These theories are here to stay. They may someday be incorporated into more encompassing unified theories, and they may break down under extreme conditions in ways we cannot now anticipate, but there will almost surely remain large domains of experience for which they reign as the governing theories. For example, any future theory describing electrons and photons must faithfully reproduce the results of QED for the description of atoms and radiation, and any future theory of strongly interacting particles must faithfully reproduce the results of QCD for the description of scaling deviations, jets, and so forth.

Much lower in this hierarchy are the grand unified theories we discussed in chapter 25. These theories have several things to recommend them. First of all, they incorporate QED, QCD, and the modern theory of weak interactions as their low-energy limit. That's good, because these latter theories are well established. Moreover, the unified theories go beyond these separate theories in two major ways. Within the unified theories, the scattered pattern of colours, weak interactions, and electric charges of quarks and leptons is organized into a convincing and coherent whole. In addition, grand unified theories successfully predict the relative strength of the various interactions.

But these successes, lovely as they are, are too few to be definitive. There is also the deep disappointment that proton decay has not been observed at the rate suggested by the simplest grand unified theories. It is therefore quite conceivable (though unlikely) that a radically different starting point might be required and that there is an alternative explanation of the apparent successes of grand unified theories. It may turn out that grand unified theories in the form we know them now—that is, as a simple generalization of QCD to more colours, with appropriate ultra-Higgs backgrounds hiding the symmetry—are not an accurate description of nature under any conditions.

In this hierarchy of belief, the ideas we have been discussing about cosmic matter-antimatter asymmetry and inflation must be ranked lower yet. Their successes are few in number, indirect, and at present very qualitative.

This situation could change. For instance, we might eventually learn enough to make intelligent guesses about the behavior of matter under the extreme conditions in the early universe, so that the precise asymmetry between matter and antimatter (that is, the precise abundance of quarks in the universe) or the precise kind of fluctuations resulting after inflation could be predicted and compared with observations. If all went well, the ideas would then ascend the hierarchy of belief.

Or it could be that by observing gravity waves, which interact so weakly that they could survive more or less intact from close to inflationary times, we would find relics of the first 10^{-34} seconds and be able to check some of our speculations about how things went.

For the immediate future, though, these ideas are probably best regarded as no more than "likely stories." Their greatest value may be that they give us a new attitude toward some of the most fundamental and enduring questions about the natural world.

Why is there more matter than antimatter? Why is the universe nearly, but not quite perfectly, uniform? Why is the average mass density close to critical? A few years ago, these questions would have seemed hopelessly inaccessible, if they were asked at all. Their answers appeared to be built into the initial conditions defining the universe. We have learned to regard them instead as <u>historical</u> questions, whose answers will appear as we work out the consequences of physical laws as they operate in the earliest moments of cosmic evolution. Once you have experienced the gorgeous view of the world available from this perspective, there is no getting it out of your head.

Let us turn from these sober warnings of scientific puritanism, to consider some wild philosophical and eschatological implications of the new ideas.

We have tried to trace the structure of the universe to the workings of extremely symmetric laws upon an extremely symmetric starting point. In the most radical version of this idea, the starting condition is perfectly empty space. Now, we might well ask, If all this is true, where did all that symmetry go? How can symmetric laws, working upon a symmetric state of affairs, lead to the rich and varied world we see around us?

We can begin to discern the elements of a beautiful answer to this profound question. The seeds of irregularities from which stars and galaxies will later grow are plausibly generated by tiny quantal fluctuations, greatly magnified in spatial extent by inflation and in amplitude by gravitational instability. But quantal fluctuations in themselves produce no asymmetries; rather, they take on all possible values in different branches of the lave of the universe. Concretely, if on one branch the universe is denser <u>here</u> and rarer <u>there</u>, there will be another branch on which it is denser <u>there</u> and rarer <u>here</u>. So, in principle, symmetry is never lost. However, it becomes very well hidden, because we recognize it only after taking an inventory of the different branched worlds in the universal lave.

Inflation expands horizons by a large amount, but not infinitely. Eventually, portions of the universe that are truly causally disconnected should become visible to our distant descendants. And so a vision of future history comes upon us.

Trillions of years have passed since the big bang. The known universe is cold and empty; stars long ago exhausted their fuel, and even the heat long trapped inside neutron stars has dwindled away. The microwave background, pale remnant of the initial inferno, has shifted further into feeble long radio waves. Somewhere in this cold and dismal darkness, there is a time capsule, containing records of the best products of a great and ancient civilization and instructions for re-creating its makers. For billions of years at a time nothing moves. Every erg of energy is precious. But every few billion years, a sensor pokes out of the capsule and briefly scans the sky. It is looking for light from beyond the region homogenized by inflation.

On one such scan, a hot spot comes into view. A new, younger, and more vigorous part of the universe has been discovered. The capsule stirs; a long journey toward rebirth begins . . .

Quest

It is no small part of the charm of physics that, after more than three hundred years of astonishing progress, there remain absolutely major and fundamental problems. Here we examine two of the biggest. One is an apparently witless and useless triplication of elementary particles. Can we decipher this teasing message from nature? Another is our inability to identify most of the mass of the universe. What is it?

In this section, we look at some "work in progress" on these problems and watch how people wrestle with the unknown.

NO FIRM FOUNDATION

When we think of Albert Einstein, we are likely to remember the white-haired elder statesman of science. Yet, even at sixteen, Einstein was already troubled by a paradox that would later germinate into the special theory of relativity. He asked himself what a light beam in empty space would look like to an observer who was riding along with it. To such an observer, the electric and magnetic fields, which according to Maxwell's theory constitute the beam, would appear frozen, static. Einstein, then scarcely more than a boy, had only a qualitative and incomplete idea of contemporary physics, but he did realize that such static fields cannot exist in empty space. According to other basic tenets of electromagnetic theory, they arise only from electric charges or currents.

Ten years passed between his initial diagnosis that something deep within the most successful theory in physics was unsound and his discovery of the cure. During all this long interval—long especially, one imagines, to a very young man with no secure prospects—Einstein never forgot his paradox. He came to realize, perhaps with fear and trembling, that to resolve it he would need to overhaul traditional ideas about space and time. He drew solace and inspiration during this period from reading Hume and Mach, skeptical philosophers who emphasized the empirical and conventional roots of all concepts used to describe the physical world.

Meanwhile, in an unrelated development, Max Planck hit upon a curious mathematical expression that described the hitherto mysterious blackbody spectrum, that is, the relative intensity of different wavelengths of light in thermal equilibrium. At first, his equation was little more than a suggestive restatement of the measurements, an empirical formula with no theoretical basis. Planck then supplied a derivation, of which Einstein says,

> Planck actually did find a derivation, the imperfections of which remained at first hidden, which latter fact was most fortunate for the development of physics. . . . This form of reasoning does not make obvious the fact that it contradicts the mechanical and electrodynamic basis, upon which the derivation otherwise depends. . . . If Planck had drawn this conclusion, he probably would not have made his great discovery, because the foundation would have been withdrawn from pure deductive reasoning. . . .

In fact, Planck had slipped the quantum of action into his not fully consistent reasoning. He had, without wholly realizing it, introduced an essential element of discreteness into the description of nature, an element alien to the theories of mechanics and electromagnetism as they then existed.

To the young Einstein, however, the imperfections were all too clear. Einstein was in the process of completing his special relativity theory, resolving at last the paradox he had discovered as a teenager. But now he was hit with a new paradox: Planck was deriving empirically correct equations from hypotheses that contradicted the principles of physics:

> All of this was quite clear to me shortly after the appearance of Planck's fundamental work. . . . All my attempts, however, to adapt the theoretical foundation of physics to this knowledge failed completely. It was as if the ground had been pulled out from under one, with no firm foundation to be seen anywhere, upon which one could have built.

It would be understandable if, after ten years of tension and struggle, Einstein had recoiled from yet another paradox. Instead, he hastened to embrace it—with results we discussed in prelude 2.

Soon after this, Einstein came upon yet another stumbling block while trying to reconcile his special theory of relativity with the theory of gravity. He became sure that his recent creation, culminating ten years of effort, was only a first step:

> This [the equality of gravitational and inertial mass] convinced me that, within the frame of the special theory of relativity, there is no room for a satisfactory theory of gravitation.

And so began, in 1908, another extended struggle, which seven years later culminated in the general theory of relativity. Of this time, Einstein writes a moving account:

> The years of groping in the dark for a truth that one feels but cannot express, the intense desire and the alternations of confidence and misgiving until one breaks through to clarity and understanding are known only to him who has himself experienced them.

From these three major struggles of Einstein's career, a clear pattern emerges. In each case, he latches on to some perceived fundamental weakness or contradiction in existing physical theory and worries over it for long periods of time—as long as it takes. He is concerned not to exploit existing ideas but to transcend them.

This restless style is not necessarily a recipe for success. Einstein did not play a major creative role in the development of physics after about 1925, although he lived thirty years more. It was not that he did not continue to work or that physics stagnated—quite the contrary, on both scores. The basic difficulty was that Einstein believed he saw difficulties in the foundations of quantum theory and that, characteristically, he wished to overhaul the theory rather than to exploit it. While his colleagues were applying quantum theory with great success to elucidate the workings of atoms, nuclei, and bulk matter, Einstein held aloof.

Most scientists are happiest when they are making clear progress, solving some perhaps small but well-defined and significant problems by clever adaptations of known techniques. Most people— perhaps all—feel acutely anxious and unhappy when they are "groping in the dark" or find themselves poised uneasily upon "no firm foundation." We must admire the courage of those rare individuals who, like Einstein, systematically seek out such situations.

30

Families

"Who ordered that?"
—I. I. Rabi, reacting to the discovery of the muon
(1947)

Protons, neutrons, and electrons are the building blocks of ordinary matter. One level deeper, the protons and neutrons themselves are made out of up and down quarks, each of which comes in three colours. We have deep and essentially complete theories of how these particles interact at all energies up to and including those accessible to the most powerful existing accelerators. There has even been substantial progress, as we discussed in chapter 25, toward unifying all the most common particles into a cohesive pattern.

All these wonderful achievements only make the *mystery of families* more embarrassing and more frustrating. For we find that each of these venerable particles, these building blocks of ordinary matter, is duplicated—no, triplicated (at least!)—in an apparently senseless way. It makes a mockery of our claims to comprehension.

The mystery started in 1947, with the discovery of the *muon,* a particle first noticed in cosmic-ray tracks. Muons are very short-lived particles that usually break down, by a weak-interaction process, to yield an electron and two neutrinos. By far the most remarkable thing about the muon is that almost all its fundamental properties—its electric charge, its spin, its lack of strong interactions, and the details of its weak interactions—are identical to those of the electron. In the five-bit colour language introduced in chapter 25, a muon and an electron are described by the same word. There is a difference, however: the muon is more than two hundred times heavier than the electron.

More than twenty-five years later, the same crazy thing happened again. Yet another pseudo-electron, this one christened the *tau lep-*

ton, was found. The tau lepton is almost twenty times heavier even than the muon (and therefore roughly four thousand times as heavy as an electron). And in the debris of particle collisions at higher energy, there may be even more replicas just waiting to be discovered. No one really knows what to expect—that's one way of putting the families mystery.

This triplication of electrons has gone hand in hand with a triplication in the quark and neutrino world. The process whereby these short-lived particles were discovered and studied is a story in itself, but we shall be content to summarize the results in the accompanying Table of Families. The most important result of all these discoveries is that the additional particles appear to fall into the same pattern as the more venerable particles, the pattern we discussed in chapter 25. This is the reason we say they fall into "families."

FAMILIES:

First	Second	Third
e (electron)	μ (muon)	τ (tau)
1/2000	1/10	2
1897	1947	1974
u (up quark)	**c** (charmed quark)	**t** (top quark)
1/200*	1 1/2	?
—	1974	not yet
d (down quark)	**s** (strange quark)	**b** (bottom quark)
1/100*	1/5*	5
—	1947	1977

Members of the three families, with the mass of each expressed as a multiple of the proton mass and with date of discovery.

Note that the mass of a quark is a slippery concept, since quarks are never found free. This uncertainty is most important for the light quarks, and so we put a * next to their estimated masses. The up and down quarks were never properly "discovered": they make up the proton or hydrogen nucleus, which goes way back. The discovery of the top quark is eagerly and confidently anticipated as higher-energy accelerators come on line.

By the way, the fact that this pattern is repeated over and over for the three families is already a significant truth. It gives us some confidence that the pattern has fundamental significance. And so our speculations about unification, which served to motivate the pattern, are very much encouraged.

Before launching into a discussion of what the triplication of families may mean, let us discuss what the other families might possibly be "good for." This discussion can be brief. The members of the higher families, aside from the neutrinos, are all highly unstable. Also, because they are heavy, it takes a large concentration of energy to create them. Rarely born and quick to perish, these particles are not found in ordinary matter on Earth, or even in stars. Their existence in the natural world is probably confined to fleeting appearances in the debris of cosmic-ray collisions—and to the early moments of the big bang, when there was plenty of energy to create them.

Sheldon Glashow, when addressing physicists whose specialty is outside of high-energy physics, draws a line between the first family and the others. "On this side are your particles. On this side are our particles." And for the most part it is true that in the more than forty years since the discovery of the muon, there have been few practical applications, or even ideas for practical applications, of the higher families. Because some of the exceptions are quite entertaining, let's take a moment to discuss them.

Muon catalysis of nuclear fusion is an idea that comes tantalizingly close to working. If it could be made to work, it would change the world overnight. When two light nuclei combine to form a heavier one, energy is liberated. The simplest possibility, and one of the most promising, is the fusion of two deuterium (1P, 1N) nuclei to make helium (2P, 2N). This sort of process is what powers our sun, and also fusion bombs. Fusion could yield a virtually limitless supply of energy, if it could be made to occur in a controlled way. It would be a much cleaner and safer way of exploiting nuclear energy than the fission processes currently used at reactors, because it would involve small amounts of material and would not produce dangerous radioactive by-products.

The difficulty in making fusion reactions go is that atomic nuclei are positively charged, so they repel one another. This barrier has to be overcome before the attractive but short-range strong force can take over and bind the nuclei together. The standard approach to fusion is simply to heat a gas of the light nuclei until they have enough energy to overcome the repulsive barrier. The problem with this approach is that a very hot gas is hard to contain. It tends to smack into the walls of whatever is holding it and to stick or cool down or both. Although the promise of fusion has been clear for forty years and although massive research efforts to use it continue, its ultimate practical utility is still not assured.

Muon catalysis is quite a different approach to overcoming the electric barrier to fusion. The basic idea is elegantly simple. When a negatively charged muon is bound to a deuterium nucleus, it effectively cancels the positive charge of that nucleus. The charge barrier disappears, and fusion can occur much more easily. In fact, it is no longer necessary, or desirable, to heat the deuterium to achieve fusion. When a fusion reaction occurs, the muon will often take up some of the liberated energy and escape, later to be captured elsewhere and repeat its work again. Such processes of muon-catalyzed fusion have actually been demonstrated; in the right circumstances, it has recently proved possible to get over a hundred fusion reactions from each muon. Although, when you balance the energy liberated against the energetic cost of creating muons in the first place, it is not clear you can win, it is remarkable how close this use of muons comes to being practical. Perhaps some clever twist on the idea will make a go of it.

(You may wonder why it helps to use muons here rather than electrons, which are also negatively charged, and come for free. The point is that muons, being two hundred times heavier than electrons, are much less subject to quantum-mechanical uncertainty in position. The "atom" a muon forms, orbiting around a deuterium nucleus, is two hundred times smaller than a similar atom involving an electron. Being concentrated nearer the nucleus, the muon does a much better job of neutralizing its charge.)

Another possible class of applications of muons is based on their unique power to penetrate large amounts of ordinary matter. By shooting muons through an object, and studying how they get absorbed or deflected, you can learn things about the material inside. Basically, you construct its density profile—in effect, taking a picture similar to a traditional X ray, but on a much larger scale. It may seem a rather exotic procedure, but when the only alternatives are digging in or breaking things open, muons may have their advantages.

Luis Alvarez and his collaborators used muons to look inside the pyramids of Egypt. They studied the ability of relatively low-energy muons, naturally occurring in cosmic rays, to get through the Pyramid of Chephren. Without damaging the pyramid in any way, they were able to show that no "hidden chambers" lurked inside.

Higher-energy muons, stretching the capabilities of current accelerators, can get through a kilometer or two of earth. Some of them will be deflected slightly by close encounters with the electric fields of atomic nuclei. By studying the pattern of deflections for a beam of muons, we should be able to reconstruct quite a bit about the density and composition of the material the beam has passed through. If recent new ideas for building more compact and cheaper accelerators materialize, muons might well become an important tool for probing the earth's crust—exploring for minerals, oil, or lost cities.

Although "practical" applications of the higher families are fun to think about and may even be important some day, they are few in number and restricted in scope. The true fascination of the families is the mystery posed by their very existence. Are they a frivolous ornament to creation—or a sign of some hidden harmony? Physicists, of course, would like to believe the latter and have developed some ingenious potential solutions of the family mystery.

One approach is to try to apply the idea of symmetry, our eighth theme, to this mystery. The basic thought here is that a hidden symmetry connects the different families, that at some level they could be interchanged without doing violence to the laws of nature. You may recall how "turning off" the Higgs field uncovered the symmetry of an electron with its neutrino, particles that differ not only in mass but also in weak colour and electric charge. To unite an electron with the muon and tau, particles whose only apparent difference is their masses, should be simpler.

Suppose that the fundamental laws of physics are fully symmetric among the families but that some background fields permeate the universe and hide this symmetry from us. The all-purpose name for such fields, whose nature and number we can only guess at, is *ultra-Higgs fields*. (We have used this term before, for the hypothetical fields that hide the symmetry of electrons with quarks.) If this view is correct, it clarifies some aspect of the family mystery immediately. We understand why the families are so strikingly similar—because we have imagined a more perfect world, in which they are interchangeable. The existence of the different families hints at a symmetry obscured from our perception, but not from our imagination.

Very well; has this idea neatly disposed of the family problem? Hardly. At such junctures, when imagination starts to fly fancy-free, I (Frank) sometimes recall my own first meeting with Rabi, at Erice in the summer of 1973. Let me preface this story by saying that Isidor Isaac Rabi is a living legend as a research scientist (he won the Nobel Prize in 1944 for his molecular beam method), founder of a great school of physics at Columbia University (several of his students have also won Nobel Prizes), and scientific statesman (chairman of the Atomic Energy Commission). Despite all that, Rabi is a kindly, open, and utterly unpretentious person. Anyway, I was startled one day when, as I was eating a lunch of bread and cheese in the central square, Rabi sat down next to me and asked me what I was thinking about. Naturally, I interpreted this to mean physics. (What I was actually thinking about was the fact that I had just swallowed a huge wad of bread and cheese whole, because he had thoroughly spooked me.) Now, this was very shortly after asymptotic freedom had been discovered, so I started telling him about how couplings turned off at short distances and how we thought this might have something to do with the strong interaction. My discussion was very abstract and theoretical, because at that moment I was utterly wrapped up in the technicalities of the calculations. Rabi was interested in the ideas, asked some simple but searching questions, and then made a remark I'll never forget. "You know," he said, "it all comes down to this: I'd really like to understand how nature works. And when you say you have an idea about it, you've got to tell me what I should do when I come into the lab in the morning."

So what can we suggest to the profoundly simple souls who would like to test with concrete experiments whether the idea of symmetry among the families is a valid one?

A promising possiblity is to look for *familons*. Familons are chips off the vacuum, related in spirit to the Higgs particles we discussed before. If the vacuum is filled not only with the ordinary Higgs field but also with additional ultra-Higgs fields responsible for spoiling the symmetry among families, then there are predicted to be many additional particles besides the ordinary Higgs particle. The additional particles are propagating disturbances in the ultra-Higgs field, just as Higgs particles are propagating disturbances in the Higgs field.

Among the additional particles are a few special ones, the familons, whose existence and properties are closely related to the basic symmetries. The basic symmetry we are considering right now is that there should be interchangeable worlds in which the different families are transformed one into the other. More precisely, such worlds <u>would</u> be physically indistinguishable—interchangeable—if the background ultra-Higgs fields were removed. Unfortunately, in practice the enormous energy that would be required to remove these fields from any considerable region of space is never available. So we cannot attain the more perfectly symmetric world. We can, however, <u>relate one imperfect world to another</u>.

The presence of background ultra-Higgs fields ruins the family symmetry. When we make one of the symmetry transformations that interchange the families, we find that the direction of the background ultra-Higgs field must also be changed. Therefore, this transformation takes us into another world. We have imagined alternative worlds, in which the ultra-Higgs fields point in other directions (just as the inhabitants of our fantasy magnetic universe could imagine alternative worlds, in which the magnetic field that permeates their universe points another way— west, say, instead of south). Because the underlying laws are symmetric, it does not require energy to move from our world to any of these alternative worlds.

It may seem that with all this talk about imaginary worlds we have strayed a long way from answering Rabi's challenge, but fortunately we can rely on nature to do whatever our imagination says is possible. Because quantal fluctuations are ever present, there is no way for a physical system to avoid exploring all possibilities eventually—"whatever is not forbidden, is mandatory." To put it another way, any transformation consistent with all basic conservation laws (including, of course, the conservation of energy) will inevitably occur spontaneously, at some finite rate.

In the case at hand, the implication is that excursions into our imaginary worlds, with families and ultra-Higgs fields rotated around, will sometimes occur in reality. The ideal fluctuation, which rotates the background field throughout the universe, will not occur in practice. Much more common, and perhaps observable, are fluctuations that approximate this ideal over a finite region of space. These will cost a finite amount of energy, but the amount can be quite small.

What can we actually observe of these peculiar goings-on in the vacuum? Well, for instance, if a family symmetry relates muons to electrons, there could be a spontaneous fluctuation in which a muon changes into an electron and a disturbance in the ultra-Higgs field. The disturbance, carrying the energy liberated in the conversion of the heavier muon into the lighter electron, will then radiate outward. This process would appear to us as a peculiar decay of a muon, into an electron and a familon. The familon is very weakly interacting; like a neutrino, it is thus conspicuous only by its absence. It takes away energy and momentum in an all but invisible form. But the electron, which is easy to detect, comes off with a precisely predictable energy, making the process as a whole quite distinctive. Very sensitive experiments are planned to look for familon emission, actually not in muon-electron conversion but in a closely related process where a K meson turns into pi meson plus a familon. (At the quark level, a strange quark is turning into a down quark plus a familon.) If familon emission occurs even once out of every ten billion K-meson decays, it will be detected within a few years.

The family symmetries we have been discussing, if they exist at all, are well hidden. It requires several kinds of ultra-Higgs background fields, in intricate arrangements, to generate a world resembling ours from a more perfect world with symmetry among the families.

This proliferation of ultra-Higgs fields leads us naturally into a wider view. Let us take seriously the idea that many ultra-Higgs fields permeate our vacuum. Then we should expect that most particles will interact with some of these ultra-Higgs fields. Those that do will acquire very large masses through this interaction. It is only the particles that avoid interacting with all the ultra-Higgs fields, and acquire mass only from their susceptibility to the relatively modest Higgs field, that we have observed. On this view, the accelerators we have, and all those we are likely to have in the foreseeable future, are not nearly powerful enough to create the vast majority of particles. Only a very few untypical ones, immune from the direct influence of ultra-Higgs fields, are available for our inspection.

To this way of thinking, then, there may be many many families and also a plethora of other particles that do not fall into families similar to those we know. In the three accessible families, we are seeing only a few accidental relics of a much larger structure. Quite possibly, for example, there are even more colours than the five we have considered before, and the difference between electron and muon is that their additional colours do not match.

In trying to reconstruct what this larger structure is, we find ourselves in much the same position as a paleontologist who tries to reconstruct some huge dinosaur from three teeth and a few footprints. The families represent a few precious fossils from a vast hidden microcosmos, whose nature at present we can barely surmise.

31

Dark Matter

The most beautiful experience we can have is the mysterious. It is the fundamental emotion which stands at the cradle of true art and true science. Whosoever does not know it and can no longer wonder, no longer marvel, is as good as dead, and his eyes are dimmed.

—A. Einstein, *The World As I See It* (1930)

There's something of a scandal in cosmology—most of the mass in the universe is missing!

This scandal is called the dark-matter mystery. More accurately, it should be called the missing-mass mystery. If the stuff were really dark—if it absorbed light—we could get a handle on it. We could determine which wavelengths of light it absorbed especially strongly, whether its distribution is smooth or patchy, and so forth. The missing mass, however, is both nonluminous and transparent. The light waves that bring us almost all our information about the distant reaches of the cosmos pass through dark matter without picking up a single clue to its identity.

If you're wondering how we can be so sure this invisible stuff is out there at all, read on, as we probe the dark-matter mystery and some exotic particles suspected of causing the problem. The dark-matter mystery is a problem in a special sense—not just an annoyance but an opportunity. There is a great discovery waiting to be made—what most of the universe (by mass) consists of.

The first hint of the mystery, which showed up as early as the 1930s, is that clusters of galaxies don't appear to contain enough mass to hold them together. The heart of the matter is the question of *escape velocity*. Consider first a local, or at least Earth-bound, example. If you fire a rocket into the air, Earth's gravitational force keeps pulling it down. It eventually falls back to Earth, unless its initial vertical speed exceeds seven miles per second—the escape velocity. Fortunately for the space program, the laws of gravitation can be used to calculate the escape velocity from Earth to a very high accuracy—the only data you need to start with are Earth's size and its mass.

The escape velocity from a cluster of galaxies can be calculated in precisely the same way. You can easily estimate a cluster's size, once you figure out how far away it is (we discussed how this is done, in chapter 4) and measure the size of its image. You can also estimate the total mass of all the stars in the cluster, by measuring its brightness—just calculate how many stars it takes to produce that much light, and add up their masses. Computing the escape velocity then becomes a simple exercise in arithmetic.

But when Doppler shift measurements are made of galactic velocities, a very peculiar result is found: the average velocity of the galaxies in the cluster exceeds the predicted escape velocity. Just as fast rockets leave Earth behind, galaxies moving faster than the escape velocity should boil out of the cluster. So the thing should be evaporating. Now, clusters of galaxies are by no means rare, and many appear to be quite old and stable, so it is next to impossible to believe that they are really just transients.

It is like a clue in a recondite mystery novel—something is clearly wrong, but what? It seems that clusters of galaxies are held together by much more than the force of gravity we anticipated. Faithful reader, you can easily guess that we will not consider fancy-free answers until the simple, conservative possibilities have been checked out. And one of the simplest, most conservative interpretations is that there is more to clusters of galaxies than meets the eye. If the real escape velocity is much bigger than the guess based on a mere counting up of the mass in stars, the clusters must contain much more mass than is visible in stars. In fact, the numbers say they must contain at least ten times as much mass in other forms.

The next clue to the mystery emerged from a law discovered by one of the earliest giants of physics, Johannes Kepler. Kepler's so-called third law describes a simple mathematical relation between the size of a planet's orbit and the planet's speed as it moves around that orbit. For example, Saturn has to travel not quite ten times as far as Earth does to complete a single orbit, but it moves so slowly that this trip takes it nearly thirty Earth years. This is very reasonable—the closer planets have to "work harder" to keep from falling in, because the gravitational pull they feel from the sun is stronger.

The same result is expected for objects revolving around any central mass. The farther away from the central mass, the slower the satellite travels. (Mathematically, the speed is inversely proportional to the square root of the radius.)

When astronomers studied the mass of single galaxies, by measuring the velocities of objects in orbit around them, they got another big surprise. For objects orbiting around galaxies, they found, Kepler's great third law fails. Gas clouds orbiting far beyond the visible region of a galaxy move as fast as closer ones. Instead of decreasing with distance, the speed seems to level off. Measurements like these have now been done for many tens of galaxies, always with the same result.

If the mass of a galaxy were distributed in the same way as its light—in other words, if its mass were concentrated in luminous objects—this behavior would be an outright violation of Kepler's law.

The alternative explanation, which we should consider carefully before tossing out a perfectly good 300-year-old law, is to give up the idea that the mass of the galaxy is concentrated where the light is. If we suppose that there is extra mass distributed in an extended halo around the visible part of the galaxy, it becomes possible to understand why the orbital speed doesn't drop off. Objects farther out will enclose within their orbits additional halo mass, whose inward tug compensates for the weakened pull of the more distant central masses.

Yet another classic mystery of cosmology is closely tied up with the problem of estimating astronomical masses: Will the present expansion of the universe go on forever? Or will it slow down and eventually reverse? As in our earlier discussion of escape velocity, the answer depends on the strength of the gravitational force tending to reduce an outward velocity to zero. The gravitational attraction depends on the amount of mass and on the average distance separating one bit of mass from another, so clearly the answer will hinge upon the average density of the universe.

For one certain *critical density*, the universe will be precisely poised between eternal expansion and ultimate collapse. The value of the critical density can be calculated from the observed rate of expansion. It turns out to be about 10^{-29} grams per cubic centimeter—roughly one hydrogen atom per cubic meter. If the average density is greater than this, collapse is inevitable; if it is less, expansion continues forever. Although the critical density is far below that of the best "vacuum" anyone has ever produced in a laboratory, it seems to be fairly close to the actual average density of the universe—so overwhelmingly do the vast nearly empty regions of space dilute the much higher local densities near stars and planets.

To satisfy our curiosity, and for the peace of mind of future generations, we would like to know more precisely how the actual density compares with the critical density.

How can the actual density be measured? One way is to add up the densities due to all the things we can detect directly: ordinary stars (they emit visible light), hot gas clouds (they emit X rays), cold gas clouds (they emit radio waves), and so forth. By the way, the microwave relic radiation contributes only about one ten-thousandth of the critical density. All these "conventional" forms of matter add up to substantially less than the critical density. The total is most likely one-thirtieth, but surely no more than one-tenth, of the critical density. When we add in mass detected by its gravitational effect, that is, by its effect on the motions of stars, gas clouds, and galaxies, a considerably larger density is found—at least one-fifth of the critical density, and perhaps more.

The discrepancy between the two ways of estimating mass is the essence of the dark-matter problem. We must consider very seriously the possibility that forms of matter invisible to all our conventional methods of direct detection make an important, perhaps even dominant, contribution to the total mass of the universe.

In favorable cases, dark matter can be "seen" by monitoring its effect on the motion of conventional matter. But there are vast spaces between galaxies where no convenient tracers exist. We have no way to measure, or even look for, dark matter in the conventionally "empty" regions that make up most of the volume of the universe.

In this murky situation, using conventional observations to try to find the true total density becomes a very dubious procedure—almost like trying to solve a murder mystery without ever leaving the scene of the crime. At this point, it makes sense to step back from the immediate observations and seek a larger perspective.

There are strong theoretical arguments that the density of the universe should in fact be very close to the critical density.

For one thing, it is an unstable situation for the actual density to be close to, but not precisely equal to, the critical density. This is because small departures from critical density get magnified in time. To see why, let's compare two universes—call them C and M. Both are expanding at the same rate, but while universe C contains the critical density of matter, universe M contains more. For both universes, gravity tends to put the brakes on further expansion. Because M is denser than C, the expansion of M gets slowed down more; its galaxies stay closer together. In other words, not only does M start out denser than C, but it also gets thinned out more slowly because it expands more slowly. The disparity between C and M—or between C and a less-than-critical-density universe L—grows with time.

This fact can be put in another and more dramatic way: for the universe to be as close to the critical density as it is now observed to be, it had to be at the critical density to better than one part in a billion at the time of nucleosynthesis, and even more accurately earlier. It would be very peculiar if the universe had been so very close to critical density all through its earlier history and were only just now departing from this pattern. So arguments from stability suggest that the density of the universe should be exactly the critical density.

Physicists' faith in this conclusion is greatly strengthened by the recent appearance of attractive theoretical ideas that predict an actual density equal to the critical density. We have in mind the inflationary-universe scenario, whose other attractions we discussed in chapter 29. One of the main predictions following from inflation is that the density of the universe should be very close to critical.

What has inflation in the first 10^{-34} second to do with the present density? Don't worry, it's not supposed to be obvious. To make the connection, it is necessary to call upon the general theory of relativity. The relevant idea is hard to prove, but fairly simple to describe. Here, we will be content to describe it.

In general relativity theory, the force of gravity arises only indirectly from the presence of a massive body. What really happens is that the presence of the body alters the geometry of space and time. Nearby objects then swerve toward it because the flat, Euclidean geometry of straight-line paths has been upset. The shortest path between two points becomes a curved line when space-time itself is curved. The important result for our purposes is that the average density of the universe can be inferred from the overall curvature of space.

It is difficult to visualize the curvature of three-dimensional space, but the essential image can be captured in a two-dimensional universe, like a sheet of rubber. The universe will be flat—Euclidean, planar—if it contains precisely the critical density of matter. If the universe contains more than critical density, general relativity teaches us, the overall curvature of the space becomes positive. That is, on large scales, space gets curved around to look like the surface of a sphere. A universe containing less than critical density acquires what's called negative curvature. On large scales, it looks like a hyperboloid or, more familiarly, a saddle.

Now, any shape gets to look flat if it is scaled up to huge proportions. That's why it's harder to see the curve of the horizon than it is to see the curve of a beach ball. So if the universe went through a period of enormous inflation, no matter what shape it began with, it would wind up looking very flat. But a flat universe, we said, must contain the critical density of mass. Inflation inevitably leads to a universe whose density is close to critical.

Observers and theorists agree, then: conventional matter is not enough. It is not enough to account for the total gravitating mass of galaxies, or clusters of galaxies. It is not enough to fill the universe to its theoretically desirable critical density. If a large amount of mass—the bulk of the universe—is some stuff that evades all existing methods of direct observation, then everything can be explained. This stuff would enable us to satisfy our theoretical yearning for the critical density and at the same time to explain the observed extra gravity.

But what could it be?

"When you have eliminated the impossible," said Sherlock Holmes, "whatever remains, however improbable, must be the truth." In this spirit, we begin our inquiry into the nature of the dark matter by eliminating as impossible the most plausible suspect. Could the dark matter be made of the same building blocks as ordinary matter—protons, neutrons, and electrons—put together in some devious way?

Various possibilities can be ruled out, case by case.

First, there are several kinds of invisible stars. A very small "star," with less than about one-hundredth the mass of the sun, would not be able to initiate nuclear burning. Such quasi-stars do exist—the planet Jupiter is a nearby example—and they would be very hard to find in interstellar space. What we need to explain dark matter is not some unusual type of star formation, however, but something so predominant that the mass of its products outweighs the rest of the universe. The limited abundance of intermediate-size objects (dwarf stars) strongly suggests that quasi-stars just don't get produced in large enough numbers.

At the other extreme of size, very large stars burn out quickly and wind up as various "dead" ashes of nuclear burning, such as white dwarfs, neutron stars, or even black holes. Black holes are, of course, nonluminous to begin with, while white dwarfs and neutron stars eventually cool down and emit no further radiation—they have no renewable energy source. Could such ashes be the dark matter? Almost certainly not, for if they did provide the dark matter, they should leave behind a telltale signature that is not found. The death of a big star, as we discussed in chapter 11, is an explosive event that hurls atoms of the heavy elements in all directions. Such explosions are, in fact, very likely the birthplace of most atoms heavier than helium. If enough big stars to account for the missing mass had exploded, there would be more heavy atoms than there are—remember, the missing mass is most of the mass, while heavy atoms constitute only about 1 percent of universal chemistry.

So much for stars. What about gas clouds? Clouds of cold gas would give themselves away by absorbing light from objects behind; they are "black clouds." Clouds of hot gas would give themselves away by emitting X rays. Such case-by-case examinations of the possibilities are instructive, but it is hard to be confident that they are exhaustive or free of loopholes. For instance, cold bricks floating in space would be very hard to detect even if they constituted most of the mass in the universe. (Bricks are harder to see than the same number of atoms spread out in a gas cloud because all the atoms inside a brick are hidden from interacting with light; only the outside layer does any absorbing. So a brick contains mostly invisible matter, just what we're looking for.) You may find it hard to take intergalactic bricks seriously as a candidate for the dark matter—it is a little difficult to imagine how such objects would ever have formed. They are mentioned here just to remind us how laborious it would be to eliminate, one by one, all possible forms of conventional matter.

Fortunately, we do not have to rely on case-by-case arguments.

There is a powerful general argument that makes it exceedingly unlikely that matter based on protons, neutrons, and electrons in any form can provide the critical density. The argument rests on the observed abundances of certain light isotopes, specifically ^{2}H, ^{3}He, and ^{7}Li—or, in more informative notation, (1P, 1N), (2P, 1N), and (3P, 4N).

These isotopes are not particularly rare, as heavy elements go; their abundance is roughly one ten-thousandth that of ordinary hydrogen. But in stars they're unstable, quickly burning down to yield ^{4}He and ^{1}H. So if the trend in stars is to diminish the number of such nuclei, we must date their formation further back, to the original big bang nuclear cooking.

As we discussed before, the main result of big bang nucleosynthesis is to stuff all available neutrons into ^{4}He nuclei. These nuclei and the excess protons (^{1}H) do, in fact, dominate universal chemistry. The nuclei of interest here—^{2}H, ^{3}He, and ^{7}Li—are minor by-products of the assembly line of big bang nuclear reactions. They will survive if the burning is not complete, just as the familiar poison carbon monoxide is produced by the incomplete combustion of gasoline.

The abundance of these nuclei depends on the density of protons and neutrons when big bang nucleosynthesis occurred. The more protons and neutrons there were, the faster all reactions could proceed and the more complete the burning. Detailed calculations show that it is possible to obtain the observed abundance of ^{2}H, ^{3}He, and ^{7}Li by incomplete big bang nucleosynthesis, but only if the total density of protons and neutrons is about one-tenth of critical.

It is suggestive, to say the least, that this value of the density, which accounts for nuclear abundances, also closely matches the density of ordinary matter actually seen in stars and gas clouds. This "coincidence" makes it seem overwhelmingly likely that conventional matter winds up mostly in conventional forms, and that most of it has been seen already. For the dark matter, we must look elsewhere.

Another dark-matter possibility, which once seemed very promising indeed, is the neutrino. For one thing, we know neutrinos exist. This may seem to be a rather modest recommendation, but we shall soon be driven to consider dark-matter candidates whose very existence is not established.

There are excellent reasons for thinking that the universe is filled with neutrinos, to an average density of several hundred per cubic centimeter. The density of this gas of neutrinos might well be a million times larger in our immediate vicinity, if neutrinos are massive and have fallen into our galaxy.

The neutrino gas is a relic, very similar to the microwave background radiation, left over from the early stages of the big bang. It was produced some one hundred seconds after the origin, about the same time that nucleosynthesis occurred—that is, at a time and under conditions we have many reasons to feel we understand well. Most physicists are confident that these relic neutrinos do exist.

Unfortunately, experimenters have little immediate prospect for detecting these relic neutrinos. The difficulty is not one of principle, but simply that their interactions are terribly feeble. To get an idea of the difficulty involved, consider the much easier problem of detecting neutrinos emitted by nuclear reactors. Ten meters away from a medium-size reactor, roughly a trillion (10^{12}) neutrinos per square centimeter pass through an observer every second. They have energies corresponding to temperatures of several billions of degrees. All this sounds pretty formidable, but the interactions of the neutrinos are so feeble that this torrent is barely detectable. To melt an ice cube with it would take an exposure time of one thousand times the life of the universe. Reactor neutrino experiments are difficult enterprises involving large apparatus and considerable patience. No effective method has yet been proposed for the more difficult task of detecting the pervasive gas of relic neutrinos. It remains an unnerving challenge to experimental physics that the universe is almost certainly pervaded by vast numbers of particles we have no effective means of detecting.

If we accept the existence of the relic neutrino gas, it follows that even a very small mass per neutrino will result in neutrinos' dominating the density of the universe. The required neutrino mass is only about one ten-millionth of the mass of the electron.

There are, despite the tininess of the possible mass involved, several ways to "weigh" neutrinos. The most straightforward is to measure very precisely the recoil velocities and energies in some process involving neutrinos and particles whose masses are known. A particularly convenient process, shown in figure 31.1, is the decay of tritium (^{3}H) into ^{3}He, an electron, and an antineutrino:

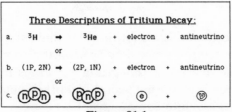

Figure 31.1

If the neutrino has a nonzero mass, the maximum energy available to the electron in this decay will be smaller than for a massless neutrino. By measuring the electron energy for many millions of such decays, experimenters hope to determine the maximum electron energy and, thereby, the neutrino mass.

The results of such experiments have set an upper limit on the electron neutrino mass; it cannot be much larger than the value necessary to supply the cosmological critical density. The experiments are, however, at least as consistent with the hypothesis that the neutrino mass is much smaller (or zero) and plays no important role in cosmology.

This method of weighing is implicitly limited to the "basic," or electron-associated, neutrino. The masses of the neutrinos associated with the other, higher families are much harder to determine. The particles that decay into these neutrinos tend to be very short-lived and hard to produce in large numbers. There are powerful but considerably less direct methods of investigating the other neutrino masses, none of which have so far given support to the idea that any neutrinos have cosmologically interesting masses. Indeed, it would require a rather peculiar conspiracy for such masses to have escaped detection if they existed.

Aside from such "technical difficulties," there are also astronomical problems with neutrinos as dark matter. There seem to be places where dark matter is just too dense to consist even of fairly massive neutrinos. The problem is that neutrinos are fermions. The repulsive identity force between fermions, discussed at some length in chapter 16, sets a limit to just how densely neutrinos can be packed together. There seems to be enough dark matter at the centers of dwarf galaxies to exceed this limit, so that at least in these cases neutrinos are not supplying the dark matter.

Here ends our discussion of neutrinos as a dark-matter candidate. On the frontiers of knowledge, there is, almost by definition, not certainty but rather a competition between different hypotheses. The hypothesis that the missing mass is supplied by the cosmological relic neutrinos has become sufficiently dubious that more speculative alternatives, involving particles whose very existence has not been demonstrated, must be considered.

If this were indeed a murder mystery, it might seem like cheating to reject all the most obvious suspects in favor of some person or persons previously unknown. Yet solutions that would be unacceptable in fiction are often, when they crop up in real life, not merely satisfactory but exciting and remarkable.

Just as the dark-matter problem was beginning to look insoluble, apparently unrelated developments in high-energy physics hinted at the existence of new kinds of particles that could constitute dark matter. Naturally, the possibility that important problems with radically different origins—from the study of matter at the very smallest and very largest distance scales—could be resolved by a single hypothesis makes the hypothesis both more interesting and more plausible.

We have, in fact, an embarrassment of riches. Two different particles—the axion and the photino—seem at present to be good dark-matter candidates. Neither has yet been experimentally proven to exist; either or both may soon be detected in experiments planned for the near future. Let's investigate them in turn.

We have already discussed the many dramatic successes of the modern theory of strong interactions—quantum chromodynamics, or QCD. We must now lay bare its persistent, and very galling, flaw.

Several years after QCD had become popular as the theory of the strong interaction, a subtle but embarrassing new possibility was noticed. It was realized that another strong-interaction coupling, which had not been taken into account before, is mathematically possible. This so-called *theta term* (Θ term), a coupling of gluons to gluons, would destroy the perfect match between the observed symmetries of the strong interaction and those predicted theoretically.

The Θ-term coupling would tend to align the spin of a neutron along an electric field, as shown in figure 31.2. This may not sound like a big deal, but it violates a symmetry that the strong interaction has always so far been seen to follow: time-reversal invariance.

Figure 31.2

Turning to the friendly realm of fantasy universes, and looking at a time-reversed T-world, we discover that electric fields stay the same but that the direction of all motions is reversed. This means that <u>spins</u> are reversed, so that the neutron in T-world is lined up <u>opposite</u> to the electric field. Because this Θ-term effect causes the laws of physics in T-world to differ from those in the real world, we say that it violates time-reversal symmetry. Experimental searches for such an effect have come up empty. Tight limits can be put on the size of any interaction that would produce them. In particular, the Θ term can be shown to have less than 10^{-9} (one-billionth) its maximum theoretical strength.

It is perfectly consistent, and adequately describes all known strong-interaction phenomena, to put the Θ term equal to zero and be done with it. But we cannot remain satisfied with this procedure for long. It's as if you tossed a penny thirty times and got heads each time. You start to suspect the thing is loaded (or two-headed).

For physicists, there is also the historical inspiration of Einstein's theory of gravity, general relativity, which he built up by trying to explain the "accidental" equality of gravitational and inertial mass. Can we do something comparable here, embedding the observed smallness of the Θ term in a larger conceptual framework, in which it appears inevitable?

The axion, the first particle ever to be named for a laundry detergent, was invented to clean up this problem in QCD.

The idea is that in a larger theory, containing QCD, the magnitude of the Θ term can become a dynamical variable, a <u>field</u> capable of assuming different values at different points in space and time, instead of a number fixed once and for all. We have talked before about such fields. In the case of the Higgs field, for example, the universe settles down to the energetically favorable situation of a high field value everywhere. What we need for a Θ field is the opposite situation, where it is energetically favorable to have everywhere a very <u>low</u> value of the field. This can be shown to occur in a wide class of theories, promising a deeper theoretical understanding of the observed smallness of the Θ term.

The attractive idea of making the Θ term a dynamical variable has as a direct consequence the existence of a new particle, the *axion*. The axion is the quantum of the Θ field, just as the photon is the quantum of the electromagnetic field.

The axion is predicted to be an exceptionally light, exceptionally weakly interacting particle—outdoing even the neutrino in both respects. But unlike the neutrino, whose fermion "unpackability" led to trouble in the center of dwarf galaxies, the axion is predicted to be a boson.

So here we have a very light boson, in many ways similar to a photon but much more feebly interacting. And, like photons, axions would have been so copiously produced in the big bang that we would expect even now to see a remnant.

During the big bang, axions were produced in a most peculiar way. Remember, the axion is the quantum of oscillations in the Θ field. If we consider how this field oscillates, it soon becomes apparent that even though the Θ field assumes its minimum energy at a very small value, it has some difficulty in settling down to this minimum. Because it interacts so feebly, it is, like a spring with very little friction, slow to come to rest.

On the other hand, the time available—the lifetime of the universe—has been considerable. It is predicted that by now Θ-field oscillations will have been damped down enough to be undetectable in T-violation experiments. Even so, considerable energy remains in the oscillations. In fact, the calculations say that just about enough energy should be stored in the axion field to account for the missing mass.

Can this striking idea be tested experimentally? Can we get the kind of access to the weakly interacting relic axion radiation that we already have to the microwave photon relic radiation? Serious efforts are being mounted to try it. The challenge is to construct a suitable "antenna" to receive this new kind of radiation, never before met in the laboratory, whose properties must be deduced from our theories of how axions couple to ordinary matter.

Right now, the most promising ideas for axion detection are based on converting the axions into electromagnetic radiation, which can then be detected by well-developed traditional techniques. Two effective "catalysts" for this conversion are strong magnetic fields and pieces of matter with many aligned spins (permanent magnets). Within just a few years, there may be definitive experimental answers to the question of whether dark matter is provided by a relic axion radiation.

Another leading dark-matter candidate is the photino. The existence of photinos is postulated in connection with a much larger theoretical speculation, the idea of supersymmetry.

Supersymmetry is a symmetry of a new and, at first sight, bizarre kind. It requires, like other symmetries, that the laws of nature in a world where the identities of certain particles are interchanged turn out to be the same as the laws of nature in our world. In chapter 25 we marveled at the audacity of the grand unified theories, which let us treat all fifteen different fermions as interchangeable. Supersymmetry goes further still, letting us interchange fermions with bosons.

Supersymmetry has many strange consequences. To ask that a world in which bosons and fermions are interchanged obeys the same laws as our world is to ask a lot. For example, supersymmetry requires that there be equal numbers of fundamental bosons and fundamental fermions, because in a transformed S-world these numbers would be interchanged. Similarly, if S-world is to look like our real world, the couplings among bosons must be closely related to the couplings among fermions.

If the theoretical concept of supersymmetry is relevant at all to the description of nature, it is well hidden. For example, in the transformed world, one would expect that the electron becomes a boson—the selectron—with the same mass. No such boson exists in our world; if it did exist, it would have been produced at accelerators and detected long ago. A broken symmetry can still be very useful, as we have seen, if it is violated in a structured way (chapter 24). It is possible that supersymmetry is spoiled only by the presence of background ultra-Higgs fields. This would mean that the fundamental equations of motion are supersymmetric but that the stable solutions of these equations—including the world around us—are not.

The power of the idea is that some (but not all) of the consequences of supersymmetry survive. In the case at hand, for instance, the spontaneous breakdown of supersymmetry does not change the brute fact that a selectron exists, but does affect its mass. It is possible to construct models in which the selectron is much heavier than the electron, too heavy to have been produced at existing accelerators.

Why are theoretical physicists so intrigued with the idea of supersymmetry, if there is little direct hint of it in the world? One reason is simply that it is fun to see whether such a bizarre idea could possibly be reconciled with the observed facts of nature.

There are other, more serious reasons too. For example, we seem to need both fermions and bosons in our description of nature. If we are ever to carry to its logical conclusion the immensely attractive idea that the world's contents and its laws flow inevitably one from the other, then some deep way of relating bosons to fermions is required. Supersymmetry offers the most promising idea along these lines so far.

Another reason, less wildly speculative but impossible to explain without using some heavy mathematics, has to do with the fact that divergent (infinite) expressions tend to show up in some of the equations of quantized field theories. We know how to deal with these infinities, but most physicists would be happier if they never occurred. It turns out that bosons and fermions contribute in opposite ways to the equations, so that better-controlled (finite) expressions can result if the boson and fermion contributions cancel. Supersymmetry enforces such cancellations.

Many hundreds of papers have been written about the construction of supersymmetric models. Many alternatives have been considered. At one extreme, it is possible that supersymmetry is so drastically broken that it leaves no experimentally accessible trace. In this scenario, the selectron, and similarly the "superpartners" of all observed particles, would be too heavy to be produced at foreseeable accelerators.

A more intriguing possibility is that the superpartners are just heavy enough to have escaped detection so far but will become evident as higher-energy accelerators become available. This idea may sound suspiciously like wishful thinking, but it is not entirely arbitrary. For if the fermion/boson cancellations mentioned earlier are to be accurate enough to explain why some things we already know about (specifically, the mass of the W boson) are as small as they are, then supersymmetry must not be too badly broken. And if supersymmetry is not too badly broken, then the superpartners cannot be much heavier than their mates. If they were, that in itself would represent a large violation of supersymmetry. So there is much interest in models where superpartners such as the selectron are not much heavier than the W boson. If these ideas are true, superpartners will be produced at large accelerators in the near future.

Remarkably, models whose supersymmetry is broken only mildly supply us with another good candidate for the dark matter. This is the *photino,* the superpartner of the photon.

The interactions of the photino follow from its role as the superpartner of the photon, according to the basic principle of supersymmetry. The photino, like the photon, interacts with electrically charged particles. But while the photon responds to electric charge without changing the identity of the particle it couples to, photinos transform electrically charged particles into their superpartners. For example, as is shown in figure 31.3, an accelerating electron can emit a photon and remain an electron, or emit a photino and turn into a selectron.

The interactions of the photino always involve at least one other superpartner. This means that, because the photino is predicted to

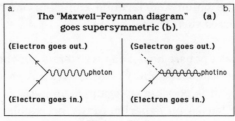

Figure 31.3

be the lightest of all "super" particles, it will be stable. It cannot give birth to other particles, since all its potential offspring are heavier than the photino itself.

Processes in the early stages of the big bang could have produced great numbers of photinos, many of which would have survived to the present. These survivors will, of course, exert gravitational pulls on ordinary matter. On the other hand, the nongravitational interactions of relic photinos are very feeble. They arise only as an indirect consequence of the exchange of virtual heavy superpartners, as in figure 31.4. In five words, relic photinos are dark matter.

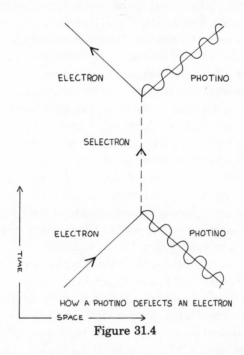

Figure 31.4

How <u>much</u> dark matter they provide depends on just how many photinos were produced in the big bang. Remarkably, it turns out that if the mass of the photino is a few times, or a few tens of times, the mass of the proton—as it is in many popular theoretical models—there can be just enough photinos left to supply the missing mass.

How will we find out whether photinos are in fact the dark matter?

In part this question is connected with that of supersymmetry as a whole. The same models that give us photinos as a dark-matter candidate predict the existence of a plethora of other new particles—superpartners for all the known particles, and more—which should all be accessible to accelerators just a little more powerful than those available today. Most of these particles are highly unstable, and photinos will be among the products of their rapid decay. As with other very weakly interacting particles, these photinos will be hard to detect directly. Instead, their existence would have to be inferred from painstaking studies of the energy and momentum they carry off.

If we knew photinos existed, we would be very tempted to believe that they provide some or all of the missing mass. Strictly speaking, though, the question of whether photinos exist and are produced at accelerators is logically distinct from that of whether a relic gas of photinos pervades the universe. It is fortunate that the relic gas itself has potentially detectable consequences.

Although the relic gas is very dilute, at best less than one particle per cubic centimeter in our neighborhood, occasionally two photinos will meet and annihilate. We will not draw a figure for this—just turn the figure 31.4 sideways! Such annihilations can yield products—including antiprotons, positrons, and gamma rays—that rarely arise from more usual astrophysical processes. The observed small flux of antiprotons, positrons, and gamma rays is comparable to what would be expected if annihilation of dark-matter photinos were the sole source. Further study of these cosmic rays may make it clearer whether conventional sources can account for them, or if there is room for a contribution from photino annihilation.

There are also exciting prospects for direct observation of the relic photino gas. Cosmic photinos will occasionally interact in ordinary matter, depositing energy. In principle, a detector could be constructed that is sensitive to this cosmic heat source. The principle is simple enough—cool a large amount of matter very close to absolute zero, and watch for occasional hot spots. We discussed this idea before, in connection with neutrino detection (chapter 16). Severe practical problems arise because of the very low rate of events and the tiny energy per event. It strains both patience and ingenuity to detect anything at all. An even more daunting problem is to distinguish what is at best a rare and tiny signal from other, less interesting sources of heat (cosmic rays, natural radioactivity, earth tremors, somebody down the hall sneezing). Nevertheless, the experimental challenge of detecting a relic background of photinos is inspiring heroic efforts. There is considerable optimism that, if they are out there, they will be found within a decade.

Axions and photinos appear to be the most promising, best-motivated dark-matter candidates. It is very exciting that experimental information about them should soon be forthcoming—but if neither proves out, what then?

Physicists are generally reluctant to contemplate the idea that the law of gravity itself may need modification. The idea runs afoul of Occam's razor and a vast amount of experience. The known laws of gravity have been used, for example, to guide space satellites on long flights and to predict the course of planets with exquisite precision.

There is one distinct possibility for modifying gravity, however, that deserves mention.

A major concept that pervades modern physics is that empty space—the vacuum—is a dynamic, highly structured entity, boiling with virtual particles and permeated by background fields. From this concept a puzzle arises: Why is gravity indifferent to all this structure—why doesn't empty space weigh? Whether you choose to describe this possibility as a peculiar form of dark matter, or as a change in the law of gravity, is a matter of taste.

It is quite easy to include a weight for empty space in the equations of gravity. Einstein did so in 1917, introducing what came to be known as the cosmological constant into his equations. His motivation was to construct a static model universe. To achieve this, he had to introduce a <u>negative</u> mass density for empty space, which just canceled the average positive density due to matter. With zero total density, gravitational forces can be in static equilibrium. Hubble's subsequent discovery of the expansion of the universe, of course, made Einstein's static model universe obsolete. Einstein quickly abandoned the cosmological constant, calling it "the greatest blunder" of his life.

The fact is that to this day we do not understand in a deep way why the vacuum doesn't weigh, or (to say the same thing another way) why the cosmological constant vanishes, or (to say it yet another way) why Einstein's greatest blunder was a mistake.

Or perhaps the vacuum does weigh, just a little bit? That would be the ultimate in dark matter.

Of course, this form of dark matter would not cluster around galaxies and thus does not explain the anomalous speed with which objects revolve around galaxies. So it can't be the whole story. Furthermore, the density associated with empty space is constant, while the density in matter decreases as the universe expands. That these densities happen to be closely comparable at present—whereas in the past matter dominated and in the future empty space will dominate—strains credulity. Nevertheless, otherwise rational cosmologists have, out of frustration with conventional matter and skepticism concerning the dark-matter candidates offered by particle physicists, been driven to consider this alternative.

The fact is that no one knows—yet—what most of the universe is made of, or why gravity pointedly ignores the vast energies locked up in ultra-Higgs fields and in the never-ending hum of the universal sounding board that permeates the universe. The final chapter in this detective story has yet to be written. You may read all about it in tomorrow's paper, or the solution may continue to elude us for a long time. Here, as on many fronts, the human struggle to understand the physical world continues, aided by powerful instruments and inspired by a rich heritage.

Success is expected, but never guaranteed.

32

Hidden Harmonies

*We came less and less to require that Love should
be enthroned behind the stars; more and more we
desired merely to pass on, opening our hearts to
accept fearlessly whatever of the truth might fall
within our comprehension.*
—O. Stapledon, *Starmaker* (1937)

People struggling to build intelligent machines have come to appreciate how long and tortuous is the path from raw perception to even the crudest useful picture of the world. Computers can easily be equipped with microscopes, telescopes, or prisms that will allow them to "see" exquisitely small details, dim distant objects, and worlds of color that human eyes could never know. But to tame this torrent of raw information, to channel it into a manageable flow, is quite another matter. No one has yet been able to build a machine that can recognize letters, or sort differently shaped and colored blocks, with anything approaching the ability of a human child.

To put it another way, the world does not present us with an obvious or unique interpretation of itself. Making sense of the world only begins with raw perception, and requires much more than passive observation.

What human brains have, and computers lack, is a set of prefabricated models of the world. Our interpretation of what's going on in the world is not the result of direct communication with nature but rather of a comparison of our raw perception with our preconceptions.

Consider the case of a very simple brain: a frog's-eye view of nature. Frogs are very efficient at catching flies—their lives depend on it. But what do they know about flies? What does "fly" really mean to a frog? In fact, a frog's ideas about flies are comically primitive. The frog will flick out its tongue, exercising its fly-catching technique, whenever it spies a dark object of roughly the right size moving through its visual field at typical fly speed. Amazingly, a real fly that *isn't* moving tempts the frog not at all. There is a rather pathetic photograph, famous among psychologists, of a hungry frog squatting immobile, while all around him dangle potentially tasty and nutritious—but motionless and therefore essentially invisible— dead flies.

So we suspect that a frog's notion of "fly" is very unlike our own. The frog, which cannot grasp that a fly is capable of standing still, is unlikely to be aware that flies eat carrion or even that flies have wings. A frog on the watch for flies relies completely on a very crude model, something along the lines of "fly: a dark spot of the right size moving at the right speed." This rough-hewn model is the frog's idea of a fly, one of the main ideas by which it interprets and makes sense of an otherwise overwhelmingly complex universe. And in the frog's natural environment, undisturbed by frog psychologists, this idea of a fly is adequate. Working with minimal neural hardware, evolution has produced a rough-and-ready system for identifying flies that really works, a system that has kept frogs fat for tens of millions of years.

Human beings, of course, are much more sophisticated. But it's important to realize that we, too, use models and that our models, too, do less than full justice to reality. Suppose some extraterrestrial intelligence were to interest itself in your concept of "fly." Looking at a fly, are you aware of the contents of its stomach? Can you analyze its chemical composition? Can you sense and interpret the pattern of neurons firing in its little brain? Of course not, even though some of this information might be useful (for instance, in deciding when to swat). But like frogs, human beings can deal with only a limited amount of information. Evolution has given us an "adequate" perception of the world—enough to ensure survival and reproduction in a variety of natural environments—and our sensory apparatus spares our brains the burden of coping with more.

Even though our senses provide us with carefully limited information, much of the power of human brains is devoted to organizing the data of raw perception. For example, light from the outside world is projected through the pupil onto the retina, a two-dimensional bowl-shaped sheet of cells. There, its presence is quickly encoded in a pattern of nerve impulses. After a few more steps of rapid but sophisticated processing, the pattern is interpreted in the framework of an *object-oriented* model of the world—as an image of objects in three-dimensional space. Such a picture is meaningful to us because it matches our model of the world. If we were to define the human visual "world" as we did the frog's idea of "fly," we might come up with something like "world: three-dimensional space populated by objects having definite boundaries, shapes, and distances." We use this model to match the patches of colors and darkness the eye sees with the objects in space the brain expects to see.

The process is so automatic that we take it for granted. But consider this: to infer the geometry of a three-dimensional scene from a two-dimensional picture of it is, strictly speaking, impossible. (Having two eyes helps, but your picture of the world certainly does not collapse when you shut one eye.) Astronomers run into this problem regularly, and they have had to come up with all kinds of sophisticated tricks to reconstruct the third dimension of the sky. Yet people in their natural environment are doing an "impossible" task all the time, converting two-dimensional retinal images into a three-dimensional scenes. They do it, rapidly and without conscious thought, by making many hidden assumptions.

It remains a lively subject of scientific research just what these assumptions are. But they very probably bear a family resemblance to the frog's assumption that "a dark spot of the right size moving at the right speed" is something good to eat. Our hidden assumptions were chosen in the process of evolution, to provide an "adequate," rough-and-ready understanding of our natural environment—the surface of the planet Earth, at roughly 300 degrees Kelvin and normal atmospheric pressure. The most important things our hungry, vulnerable ancestors had to deal with—notably, potential food, predators, and each other—are all quasi-permanent, solid objects at roughly human-size scales, located in three-dimensional space.

So even though "out there" no one has found a built-in meaning to the universe, "in here" our brains seem to have a few built-in ways to make sense of perception. Now, of course, organizing raw perception into a world of objects is only the beginning of human thought. It is the first step and is likely to be pretty much the same in its fundamentals for all people. Much more is required, however, to form a satisfactory picture of the world. For example, using the past behavior of some object to predict what it's likely to do in the future is a little like adding the idea of time to your impossible three-dimensional picture, to create an even more impossible fourth dimension. More assumptions are made; perception is sculpted in other ways and fitted to other models. Much of this is done below the level of conscious thought, though—as every driver knows—the subconscious learns a great deal from experience.

In the end, we filter reality through our preconceptions and, like the frog, see what we are prepared to see. The poverty of a frog's sensory and mental apparatus is painfully apparent to us. We recognize that it sets crippling and inescapable limitations upon the frog's appreciation of the world. But is our own position really qualitatively different? Don't inescapable human limitations cripple our own ability to appreciate the world?

There is reason to think not—to think that our species will be able to transcend any bound upon its capabilities—for we human beings, despite our many limitations, are enormously flexible. We can learn to make use of our innate capacities for purposes far different from those for which they were originally evolved. Legs evolved for trekking over the savannas of Africa can be taught to glide on ice skates. Muscular skill and coordination evolved for hunting and escaping predators can be used for baseball or belly dancing. Minds evolved for survival in a varied and often hostile environment can be used to write novels or to solve equations in quantum mechanics. Our limitations seem to be eminently escapable. By learning, building tools, and cooperating, humans manage to do many impressive and unnatural things.

With the blessing of flexibility comes the happy burden of making choices. The mental resources that make us human may not be infinite, but they can be deployed in a bewildering variety of ways. Some of our "choices" are unconscious, or involuntary. For example, although every normal human child learns a language, the particular language it learns depends on the culture into which it is born. Each language brings with it some characteristic emphases and modes of thought. Children in temperate climates learn to recognize generic "snow"; eskimo children learn to recognize and name many varieties of snow, each with different implications for the state of the world (is spring at hand, or a fierce blizzard?) and rich emotional connotations. These childrens' appreciations of snow may well be as different as the frog and human appreciations of a fly.

Such ideas and examples suggest a view of the nature and function of science that we think you will find accords beautifully with the account of physics you have just read.

The concept of "objects in space" is probably common ground for all of us, a very basic way of making sense of the universe. As we get older and more sophisticated, we look for world models that offer more predictions. This is a good idea: like frogs, we can see only what we are prepared to see. Unlike frogs, we get to choose what we are prepared to see.

Science, and in particular physics, offers one model of the world— the model we have been discussing in the preceding pages. Science is not alone in its claim to offer us a system for making sense of the universe. Every culture, every religion, offers its own world model, something to tell us what we should expect to see "out there."

There is something very special about the world model that science provides, something that (to us) makes it especially legitimate.

The point here is not that science is right and other ways of looking at the world are wrong. On the contrary, the main problem with many nonscientific world models is the vigor with which they insist upon their rightness. Once a world model claims to be completely right, it is no longer open to any changes. If, for example, you "know" that the universe was created in seven days a few thousand years ago, you must disbelieve any measurements suggesting a different age for the universe. It can do you no good, and it may do harm, to examine new data or ask further questions.

Closed systems can be comforting, but they are limited. When you commit yourself to a closed system, you are taking the road of the frog. A closed system may be adequate, but it can never strive toward perfection. It's not the best we can do.

Neither is the extreme "open-mindedness" that slides into "empty-headedness"—the idea that we can never really know anything. What is that but one more self-contented, frog-like system?

Science, on the other hand, is truly the opposite of a closed system. Scientists spend their lives constructing models that try to make sense of the world, but their focus must be on the world, not on the models. The whole idea of science is really to listen to nature, in her own language, as part of a continuing dialogue. In this dialogue, the scientist plays the role of straight man, asking questions in the form of experiments, but nature's answers are what the audience has come to hear. However much we may like an idea, if it can't be tested by experiments, it isn't science; and if the experiments don't live up to its predictions, it is wrong.

The predictions of traditional mythologies have not fared well. Despite hundreds of years of intimate dialogue with nature, scientists have yet to hear from a single extranatural willful agent. Nor does the underlying premise, that human beings play a pivotal role in the cosmos, ring true. We are dwarfed by the scale of the universe, we seem to occupy an undistinguished position within it, and the power of all our efforts is nothing by comparison with the output of the smallest star.

The straightforward interpretation of nature's message is that we human beings are rather accidental, esoteric details in the expression of her grand themes. We should look for joy not in wishful thinking about the message of nature but in the dialogue itself, and in the realization that we partake of her many hidden harmonies.

"To an astronomer," someone once said to Einstein, "man is nothing more than an insignificant dot in an infinite universe."

"I have often felt that," replied Einstein. "But then I realize that the insignificant dot who is man is also the astronomer."

Dear reader, you have faithfully accompanied us through this book, trying to listen to nature speaking in her own language. Together, we have followed on the track of physicists exploring far beyond the normal human environment to the most extreme situations—the tiniest bits of matter, or the largest and most distant; the highest concentrations of energy, or the coldest possible conditions. And together we have heard that nature can sing some strange and unfamiliar songs.

In coming to appreciate these songs, we develop a heightened perception. The rough-and-ready perception with which evolution has supplied us is leavened by an admixture of our own creation, so that we come to perceive strange but simple worlds at the foundation of our own.

We come to perceive that ordinary objects, the ones our usual rough-and-ready perception is equipped to handle, are complicated secondary manifestations of the fundamental themes—stable patterns of laves, tiny disturbances in universal oceans dancing together in choreographed arrays.

We also come to perceive, in the structure and content of the world, traces of its profoundly simple but alien past, for this structure and content are, at least in large part, the inevitable consequence of the play of nature's fundamental themes in history. What links the past—out to its farthest reaches—with the present is not the persistence of specific objects but the lawful procession of metamorphosis.

The worlds opened to our view are graced with wonderful symmetry and uniformity. Learning to know them, to appreciate their many harmonies, is like deepening an acquaintance with some great and meaningful piece of music—surely one of the best things life has to offer.

Notes

xiii. The lines are from T. S. Eliot, "Little Gidding," in *Four Quartets* (New York: Harcourt, Brace, 1943), 39.

5. The association between specific colors of spectral light and waves of definite wavelength will be discussed at length in chapter 12.

12. This passage and quotations from Galileo and Kepler throughout our book are taken from Arthur Koestler, *The Sleepwalkers* (1959; New York: Grosset and Dunlap, 1963).

20. Avogadro's number is named for the Italian physicist Amedeo Avogadro (1776–1856). In 1811, he argued that equal volumes of any kind of gas, at the same pressure and temperature, will always contain the same number of molecules. His proposal was not generally accepted until about fifty years later, when another Italian, Stanislao Cannizzaro, used it to explain the rules governing chemical reactions (the law of mass action). The *value* of this number of molecules, which Avogadro identified as a universal constant, was not determined until the twentieth century. Knowing its value is equivalent, if you think about it, to knowing the weight an individual molecule.

21. Brownian motion is briefly defined and discussed in prelude 2. A much more extensive, but quite accessible and entertaining, discussion of this and related topics is given in P. W. Atkins, *The Second Law* (New York: W. H. Freeman, 1984).

27. The photoelectric effect is described in chapter 13.

31. From his pioneering studies of double stars—twin stars, often of very different brightnesses, that orbit around each other—Herschel knew that stars are not all equally bright. He deliberately oversimplified. Why? Because he desperately wanted to satisfy his curiosity regarding the shape of the system of stars, and only by oversimplifying could he hope for an answer, however imperfect, in his lifetime.

32. The images of gas clouds within our galaxy and of external galaxies both appear as indistinct smudges in small telescopes and were classed together as "nebulae" in the astronomical literature into the early twentieth century. The term *galaxy* was used solely to refer to our own Milky Way (indeed, the Greek *galaxias* translates as "Milky Way"). Today the accepted usage is to reserve the term *nebula* for a gas cloud, while *galaxy* refers to a class of much larger, more distant objects, each containing many millions of stars.

33a. Most galaxies are elliptical; many are, like our own Milky Way, flattened disks with spiral arms. A small percentage of galaxies, perhaps one in twenty, has no simple regular form. The irregular galaxies probably result from the disruption of an initially regular structure by tidal forces, caused by the gravity of other, nearby galaxies. For example, the nearest galaxies to us, the Small and Large Magellanic Clouds, are irregular galaxies. Comparatively small galaxies, they are distorted by the pull of the much larger Milky Way.

33b. Recent observations have suggested that galaxies may be concentrated in thin shells, tracing a sort of foam or sponge that borders void regions. These findings may indicate a new level of structure in the universe, but they do not really challenge the basic fact of large-scale uniformity. It's sponge all over.

39. Any two isolated astronomical bodies, that is, bodies whose motion is not perturbed by forces special to their local surroundings, will be moving apart. Because each part of a ruler exerts powerful cohesive forces on its neighbors, ordinary rulers are definitely *not* included in this framework. Galaxies are.

40. To keep the discussion simple and concrete, we have supposed that the ruler has an end. Strictly speaking, this ruins the uniformity—the point at the end is definitely different from the others. None of the following arguments are affected by this bit of sloppiness. If so inclined, the reader is invited to imagine an infinite ruler.

42. Time-reversal invariance is not exactly true for all the laws of physics. For our immediate purposes in the present chapter, time-reversal invariance is close enough to the truth to be a convenient fiction. We shall return to this topic in chapters 18 and 27.

45. Strictly speaking, the neon inside a neon light is not in thermal equilibrium, so one must be a little careful in speaking of its temperature. For present purposes, the essential point is that the amount of energy available to pull electrons out of atoms is the same as what exists at a high temperature.

51a. The stability of nuclear decay rates is one aspect of the "all or nothing" nature of quantum mechanics, which will appear as our sixth theme. Small forces on a nucleus do not produce small changes; they produce no changes at all. In our context, all forces short of collisions disrupting the structure of the nucleus are negligible.

51b. Our notation for the composition of nuclei is very simple: for example, uranium (92P, 146N) contains 92 protons and 146 neutrons. *Isotopes* are nuclei containing the same number of protons but a different number of neutrons; they are chemically identical. Nuclei are discussed much more fully in theme 4.

To return to the matter at hand, uranium (92P, 146N), or ^{238}U, is the most common form of uranium, while uranium (92P,143N), or ^{235}U, forms slightly less than 1 percent of all natural uranium. ^{235}U is the isotope used to make atomic weapons. One main difficulty in manufacturing such weapons is to extract this rare isotope, which is chemically identical to, but much rarer than, the common form. Elaborate and expensive methods are required. It is a horrible speculation that civilizations formed earlier in the history of the galaxy would have found it much easier to manufacture atomic weapons (since ^{235}U was more abundant) and to destroy themselves.

52. Naturally, a star does not have written on it what it weighs, or whether it is burning hydrogen. We rely on the theory of stellar structure to relate these properties to the observable features of stars (principally their spectra).

63. The small difference in mass between the whole nucleus and the sum of its parts reflects the binding energy of the nucleus. According to Einstein's famous relation between mass and energy, the difference in mass is the energy that would be liberated in forming the nucleus from its individual protons and neutrons, divided by the square of the speed of light.

64.   It was not the decay of free neutrons that was first observed, but the decay of neutrons sitting inside large atomic nuclei. We will here not attempt to do justice to the real, quite complex history, since that would conflict with our aim, which is to present important themes in their simplest logical form.

66.   The idea of trajectories should not be taken too literally, as we shall see in our later discussion of quantum mechanics. Speaking more precisely, it is a probability distribution for the energies and angles of the decay products that is calculated.

67.   The "inevitability" of these other transformations will be justified in chapter 18.

86.   More accurately, Helmholtz and Kelvin estimated that the sun could maintain its present temperature for only a few million years.

89.   The following account applies only to certain supernovae (Type I), and even for these it is slightly oversimplified. Stars can be supported, after their fuel is exhausted, by a peculiar effect described in chapter 16, becoming white dwarfs. But when too much mass accretes on a white dwarf, it explodes.

90.   During helium burning, some helium (2P, 2N) combines with the resulting carbon (6P, 6N) to produce oxygen (8P, 8N) and, in turn, with the oxygen thus formed to produce neon (10P, 10N). These isotopes are therefore also predicted to be important components of highly evolved stars; and they are among the most common of heavy nuclei.

91.   The causes of supernova explosions are complex and not entirely understood. In fact, the triggering mechanisms almost surely differ for different supernovae. Nuclear detonation, discussed here, is one important mechanism, but not the only one.

95.   The rate at which energy is lost is proportional to the surface area, while the rate at which energy is generated is proportional to the volume. Thus it takes a more powerful energy source to keep a small planet going. For the same reason, to "eat like a bird" really means to consume a large fraction of your body weight each day, just to replace the energy you lose as heat.

104.   The molecules that originally hit the sphere are not at all the same ones that eventually hit our eardrums. When energy is transmitted by a wave, what is passed along is a state of excitation, not a physical object. One consequence is that sound travels much faster as a wave than it would if its information had to "ride" on individual molecules. Typical air molecules have large velocities at any one moment—comparable to the speed of sound—but they do not get very far very quickly, because they are always colliding and changing direction. This is why if someone drops a bottle of perfume, you will hear the bottle shatter long before you smell the perfume. Sound, the excitation of the air produced by the shattering, travels much faster than do the molecules that excite our sense of smell.

122.   The quotations are from John von Neumann, *Mathematical Foundations of Quantum Mechanics,* trans. Robert T. Beyer (Princeton: Princeton Univ. Press, 1955; originally published in 1932), and Werner Heisenberg, "The Representation of Nature in Contemporary Physics," *Daedalus* 87, no. 3 (1958): 99.

124.   We are describing here what is, for reasons that will soon be clear, called the "many worlds" interpretation of quantum mechanics. This interpretation was championed by H. Everett III in his 1957 Ph.D. thesis, directed by John Wheeler.

132.   Earth, of course, stays in orbit around the sun indefinitely, even though it is attracted toward the sun. An electron orbiting close by around a proton, however, is in a much more precarious state (because the forces are much stronger and the accelerations much larger). It can be calculated that the electron would radiate light and spiral down to the nucleus in a small fraction of a second, were it not for quantum mechanics.

147. ³He nuclei possess spin, so to make them identical we must be sure that they are spinning in the same direction.

148. The logic behind this even-odd rule is straightforward. Suppose we set up the beam experiment to shoot identical <u>bundles</u> of fermions at one another. When identical bundles interchange trajectories, you can imagine each fermion in turn giving its fermionic instructions: "Subtract that path!" But when you <u>subtract the subtraction</u>, that is just adding.

For instance, consider beams of deuterons, which are made of one proton and one neutron stuck together. We want to know if the two possible trajectories for deuterons add (constructive interference) or subtract (destructive interference). Interchanging the deuteron trajectory, we are changing both a proton and a neutron at the same time. Since the proton is a fermion, we would get (other things being equal) canceling contributions from the two trajectories. But other things are not equal—the neutron is a fermion too. Therefore, we must change the sign again, and then instead of cancellation there is constructive interference. If you think it through, you will see that our odd-even rule for composite particles is the only logical one.

152. You may well ask, what is it that's vibrating in the electron lave. It is easy to visualize how the shape of a string or plate vibrates in time—but a wave of probability? The next paragraph gives an honest answer but requires some familiarity with functions and complex numbers. We will not rely on such familiarity elsewhere.

The electron lave is described mathematically by assigning a complex number to each point in space at every moment of time. In other words, it is described by a complex function of space and time, called the wave function. The square of the magnitude of the wave function at a given time and a given point in space is the probability (per unit volume) that if you measure precisely <u>then</u> you will find the electron precisely <u>there</u>. What "vibrates" is not, however, the magnitude but the *phase* of the wave function. When the electron is in a definite mode, the phase of the wave function varies periodically. This means that the *probability* does not change at all. It is only a rather less tangible thing—namely, the phase of the wave function— that oscillates.

161. Popular science books often claim that the theory of relativity showed that the either doesn't exist. Not so. What the special theory of relativity did show is that people had mistaken ideas about the properties of the ether. Some of the intuitions you bring from experience with other fluids are wrong for the electromagnetic fluid. In particular, the electromagnetic ether has the very surprising feature that it looks the same to a stationary observer and to an observer moving at constant velocity. This behavior is unlike that of fluids we ordinarily run into—generally, if you run into them, they let you know it; you feel their breeze and are slowed by their drag. But although the electromagnetic field, or ether, is a peculiar sort of fluid, physicists would find it hard to get along without it.

162. Forgive us, Sir Isaac! Newton himself put forward his theory of gravity acting at a distance through empty space without dogmatism; in response to criticism from Leibniz, he was even driven to write,

> [T]hat one body may act upon another at a distance through a vacuum, without the mediation of anything else, by and through which their action and force may be conveyed from one to another, is to me so great an absurdity that I believe no man who has in philosophical matters a competent faculty of thinking can ever fall into it.

But after his theory of gravity continued its perfect record of success for over a century, its style of explanation hardened into a sort of scientific dogma. His followers, in other words, became more "Newtonian" than Newton himself ever was.

163a. We have just discussed why a field description is practically unavoidable in discussing influences that travel through space at finite speed. Now, according to the theory of relativity all influences travel at a finite speed, limited by the speed of light. This holds, in particular, for the effect that the creation of an electron has on the rest of the world. Thus the creation of electrons must be described by a field.

163b. There is a certain poetic pleasure in devising different metaphors for the abstract realities of the physical world. In addition, such activity serves an important scientific purpose. Often the most difficult, and certainly the most critical, moments in a research scientist's career come when one project is completed and it's time to decide what to do next. At such moments, metaphors come into their own. They suggest new directions—new questions.

For example, our metaphor of different fields as describing chemicals in a universal ocean suggests one set of questions: Are the particles we know actually made of still smaller particles, the way chemicals are made from atoms and ultimately from electrons, protons, and neutrons? Can their reactions be understood as rearrangements of these more fundamental components? Can the universal ocean boil, or freeze?

Another set of questions is suggested by the universal sounding board: Is space-time embedded in a larger number of higher dimensions, so that the metaphorical vibrations of the sounding board are real vibrations in a larger space? What is the sounding board made of? Are its resonances—the specially favored vibrations, corresponding to the different kinds of fundamental particles—modes of vibration of small substructures inside? Might these substructures themselves be extra dimensions, curled up? Might there be defects—cracks in the universal sounding board—and how would they manifest themselves?

Metaphorical thinking is not a replacement for hard scientific theory, but neither is it its opponent: the metaphors both stimulate new questions and help us comprehend and enjoy their answers.

178. Anderson's original pictures were taken in a photographic emulsion, whereas figure 18.1 is a bubble chamber picture. There is no important difference in principle between the two kinds of picture, and bubble chamber pictures are a little clearer, so we chose to use one.

180. Actually, we will be talking about the logic of the world, not specifically about the logic Dirac himself followed. The path that led him to the positron was based on related but different, more mathematical reasoning.

187. The rotations we speak of here are very peculiar; they can involve making the time an imaginary number.

188a. Or is this perhaps just a local prejudice—might distant stars and galaxies be made of antimatter? It seems most unlikely. For a discussion of the evidence against antimatter in the universe, see chapter 27.

188b. A more precise version of the idea that we should begin with pure energy is that we should begin with thermal equilibrium at a very high temperature.

190. Many entries in the Rosenfeld Tables represent different ways of putting together three interacting quarks (or a quark and an antiquark), rather than associations of different numbers of quarks. In this sense, the tables more closely resemble the spectra of a few atoms than the periodic table of elements.

196. We infer that quarks must be fermions rather than bosons because protons—made from three quarks—are known to be fermions.

198. We are cheating history a bit here. The historical path along which the dynamics of colour and the existence of gluons came to be understood was not this sort of playing with analogies. Understanding came through a more subtle and indirect analysis of some surprising experimental results, as we will discuss in the next chapter.

241a. There is a standard, but possibly confusing, usage we should warn you about. The term *field* is commonly used for two different concepts. One is the active medium or sounding board we discussed earlier. In this sense, the magnetic field, for instance, always fills space. The other refers to the momentary state of the field, that is, its strength, as in "the earth's magnetic field points north and is about a tenth of a gauss." It's as if the same word were used for "sounding board" and "shape of sounding board." Once you're alert to this dual usage, it won't confuse you, because the context always makes it clear what's intended.

241b. It is possible to make mistakes by taking this naïve picture of the spin of the electron too literally. We won't. This footnote is only to forestall an outcry from the learned.

243. The unfamiliar members of this list make their appearance later, especially in chapter 30.

263. The interior of Cheyenne Mountain has been hollowed out to create a well-protected electronic "command center" into which our fearless leaders will retreat in case of nuclear war.

268. For historical reasons, in the scientific literature you will usually find a quantity called baryon number, rather than quark number, discussed. Baryon number is simply one-third of quark number. It was defined that way because the baryon number of a proton was taken as the unit. When this definition was made, people didn't know about quarks.

282a. This is not to say that size is the only factor that determines intelligence. Sheep and whales have brains larger than ours. Miniaturization of the individual modules can allow more to be crammed into a fixed space. Today's microprocessors, roughly the size of a dime, can accomplish tasks that were beyond a building full of vacuum tube technology thirty-five years ago. But it is quite plausible that in a biological context the organization of modules is more difficult to change than their number, and this puts size at a premium.

282b. The quotation is from *Last and First Men* (1931).

306. The history of the discovery of the muon is confusing. Cosmic-ray tracks that we would now identify as muon tracks were seen as early as 1936. But because muons have nearly the same electric charge and mass as π mesons, the distinctions between them—most notably their strong interactions, lifetime, and spin—were not appreciated until experimentalists pinned down these properties in 1947. So we say muons were noticed in 1936 but discovered in 1947.

321. If you count the heavens, star by star, dwarf stars are the most common type. But because they are small, they do not make a dominant contribution to the total mass in stars.

Index